THE ARGUMENT
CREATIONISM vs. EVOLUTIONISM

By
Wilbert H. Rusch, Sr.

Creation Research Society
Monograph Series: No. 3

Creation Research Society Books

The Argument – Creationism vs. Evolutionism
By Wilbert H. Rusch, Sr.
Edited by George Mulfinger, Jr.
Cover Design by Suzanne Rusch

ISBN: 0-940384-04-3

© 1984 Creation Research Society
All rights reserved. No part of this book may be reproduced in any form or by any means without permission in writing from the publisher.
Second printing August 1988

Please Note

The quoting of various individuals and sources in this work is not to be read as necessarily denoting approval of the various ideas and philosophies associated with them. The author, in fact, vehemently disagrees with, in other areas, certain individuals and sources quoted herein with approbation regarding the particular matter under discussion.

Quotes from various sources and individuals named herein apply in many cases only to the time of the actual reference given. No doubt some views have changed since the time of reference.

Printed in the United States of America

INTRODUCTION

In the last two decades, the subject of origins has become an increasingly active topic. The arguments, both pro and con, on the question of the origin of the earth and life in terms of creationism or evolutionism are varied and many. As membership secretary of the Creation Research Society, particularly the past three years have brought the writer hundreds of queries relative to some phase or other of this whole question. With such a definite need for more information on these matters, this paper was begun, based on the writer's responses to the most commonly asked questions relative to the topic of origins. These responses were then combined into a more or less cohesive whole, and from that point on, like Topsy, the paper just grew.

Obviously the nature of the questions received dictated the original direction of the paper. As it developed, further input was sought from various individuals who were involved on both sides of the argument. As more and more critiques were received, the paper was revised and enlarged numerous times. Further submissions of criticisms, suggestions, and encouragement have brought the resulting paper to this stage. Whether one agrees with the result in whole or in part, it can still serve as a basis for a more precise examination and discussion of some of the arguments, both pro and con, on the subject of origins. Finally it should be pointed out that of necessity the author realizes the size of this paper precludes exhaustion of the topic.

The following acknowledgments of special help and encouragement are made with appreciation and pleasure: Drs. Wayne Frair, E. Edward Hackman, George Howe, John N. Moore, and Alan Price. All have read the manuscript at various stages of its development. An additional source of much encouragement and help came from my sons, Dr. Frederick Rusch and Mr. Wilbert H. Rusch, Jr. Above all my wife Marge has given me help with corrections, proofreading and the constant encouragement, without which I might never have finished this paper. However, all errors are solely the responsibility of the author.

Wilbert H. Rusch, Sr.
10/1/83

CONTENTS

		page
	INTRODUCTION	iii
	TABLE OF CONTENTS	iv
I.	BEFORE DARWIN	1
	A. The Ancient World	
	B. The 4th to 18th Centuries	
	C. The 19th Century	
II.	THE ADVENT OF THE *ORIGIN*	6
	A. Charles Darwin	
	B. The *Origin*	
III.	THE AFTERMATH	9
	A. The Acceptance	
	B. The Effect on Religion	
	C. The Scopes Trial	
	D. The Darwin Centennial	
	E. The BSCS and Its Work	
IV.	TERMS AND ASSUMPTIONS	15
	A. Creation-Science and Evolution-Science	
	B. True Creation-Science?	
	C. The Law of Priority	
	D. Early Views of Science	
	E. Science	
	F. Two Aspects of Scientific Inquiry	
	G. Speculative Scientific Inquiry	
	H. Uniformitarianism	
V.	EVOLUTIONISM	23
	A. Definition of Macroevolution	
	B. Early Arguments for Macroevolution	
VI.	CREATIONISM	26
	A. Defining Creationism	
	B. Is Creation Religious?	
	C. Scientific Status of Creationism and Evolutionism	
VII.	THE LIMITS OF VARIATION	30
	A. Linnaeus' Original Position	
	B. Degrees of Variation	

VIII.	MACROEVOLUTION	32
	A. The Continuum of Life	
	B. The Loyal Opposition	
	C. Macroevolution Defined	
	D. The Fate of Early Arguments for Macroevolution	
IX.	READING THE FOSSILS	39
	A. The Paleontological Evidence	
	B. The Situation Today	
	C. The Return of Paley	
X.	FOSSIL MAN	42
	A. Human Origins	
	B. Human Variation	
	C. *Homo* Characteristics	
	D. The Human Fossil Evidence	
XI.	THE LATEST FINDINGS	47
	A. Lake Turkana Finds	
	B. 'Lucy'	
XII.	THE AGE OF THE EARTH	50
	A. The Age of the Earth	
	B. The Weaknesses of Radiogenic Dating Methods	
	C. The Evidence for a Young Earth	
XIII.	CONFLICTS AND CONTRADICTIONS	56
	A. The Creationist Parent's Dilemma	
	B. The Results in the Public School	
	C. The Level of the Problems	
	D. A Possible Solution	
XIV.	CONCLUSIONS	59
XV.	REFERENCES	60
XVI.	APPENDIX A: Bibliography	64
XVII.	APPENDIX B: Quotes	66
XVIII.	APPENDIX C: Creationist Organizations	69
XIX.	APPENDIX D: Humanist Manifesto II	77
XX.	APPENDIX E: Statement Affirming Evolution	85
XXI.	INDEX	87

I. BEFORE DARWIN

The Ancient World

The average person imagines that the existence of the theory of evolution dates back to about the appearance of Darwin's *Origin of Species*. Yet, evolutionism, or transformism, as it is also known, is a concept that seems to have made its first documented appearance in Greece about 600 B.C. There it was first promulgated in its simplest form by Thales of Miletus (640-546 B.C.). He was followed by a number of later Greek philosophers who continued the development of the concept in more detail. Most prominent among these were Plato (427-347 B.C.) and Aristotle (384-322 B.C.). Some of the later Greek philosophers made slight contributions to the advancement of the concept of transformism, but essentially it came to a standstill with the decline of Greek civilization.

It is amusing to note that secular historians generally assume that the earliest creation account is found in the Babylonian literature, and that the Genesis account is then derived from that account. However, it is just as reasonable to assume that the Genesis account came first, and the Babylonian is a later distorted and garbled version of the former. A relatively recent find, discovered about 30 miles south of Aleppo, Syria, at Tell-Mardikh, would seem to support that view. This was the discovery of royal archives, inscribed on over 15,000 clay tablets found at the site of ancient Ebla. These tablets surprised scholars by indicating that what had been considered as only a first millennium oral tradition, now was apparently shown to exist as a third millennium written document. Further it showed the existence of a pre-Biblical Hebrew.[1]

The Hebrews, on the basis of the book of Genesis in their Old Testament, had no doubt as to what the origin of the earth and its inhabitants was. It is a consistent thread that runs throughout the Old Testament showing that the Hebrews trusted and relied on their Genesis account. Furthermore, it should be pointed out that any serious reader of the New Testament shortly becomes aware that Christ accepted the Old Testament account of the creation as factual. Now we find that particularly in the last decades it has been demonstrated that this reliance on the veracity of the Old Testament has not been misplaced.[2]

In fact, substantiations of the Old Testament record have become rather numerous. The renowned ancient near east scholar, William

---

[1] Giovanni Pettinato, "The Royal Archives of Tell Mardikh-Ebla," *Biblical Archaeologist*, XXXIX, 2 (May 1976), pp. 44-52.

[2] Clifford Wilson, *Rocks, Relics and Biblical Reliability*, Probe, Richardson, Texas, 1977, Chap. 13.

Foxwell Albright, towards the close of his life, was moved to announce that he considered the Old Testament as being historically correct. He wrote that the Table of Nations found in Genesis 10 "remains an astonishingly accurate document"[3]

The Romans do not seem to have furthered the development of the theory of evolution to any great extent. In fact the only known Roman who seems to have contributed to the furtherance of the theory in any way was Lucretius (98-55 B.C.). He produced a six-volume work, *De Rerum Natura*. The problem of origins was dealt with in the fifth book of this work, where he traced the development of plant and animal forms from the mother-earth as he saw them. He held that many forms of life existed and then died out, while many others survived through the protection of craft, courage, and/or speed. He supposed that man developed out of a primitive, hardy, beast-like condition into the humans of Lucretius' day. That seems to have been the end of the development of evolutionism for more than a millenium.

The 4th to 18th Centuries

Contrary to many opinions expressed today, neither Augustine nor Thomas Aquinas supported an evolutionary theory. Augustine (A.D. 354-430) clearly points out that in his opinion matter was created actually, while living organisms were created both actually and potentially. In using the word 'potentially', he apparently was referring to procreation, i.e. future generations. However, Augustine did speculate that the seven days were not days as we know them, holding that these days were beyond the power of men to know or to say. Interestingly enough, he was very definite in saying that at the time of Creation, "neither did He (God) cause them to be propagated from other individuals, but He called them into being several at once".[4]

Thomas Aquinas (A.D. 1225-1274) reported the varied opinions of his contemporaries without necessarily indicating his own beliefs on these matters. Aquinas felt that since Aristotle was the greatest student of nature, his works contained all the knowledge necessary to human existence. Therefore man did not have to waste his time in checking the knowledge that Aristotle had presented. In his opinion, man was better occupied preparing himself for life in the hereafter. It is this opinion of Aquinas that has been misread as demonstrating that he supported evolutionism.

Many who report on the development of evolutionary theory

---

[3] William F. Albright, *Recent Discoveries in Bible Lands,* Funk & Wagnalls, NYC, 1955, pp. 70ff.

[4] Augustine, *City of God,* Modern Library, Random House, NYC, 1950, p. 406.

through the course of history, include the father of taxonomy, Linneaus, (A.D. 1707-1778) as a supporter of the theory. It is true that he was the originator of the binomial system of nomenclature, also that he developed the classification scheme known as the biological hierarchy. Yet without question, Linneaus was a creationist. In his monumental work *Systema Naturae,* Edicio Decima, 1758, he even made a definitive statement on the matter of the units of creation. It was his considered opinion, resulting from all of his observations up to that time, that the original created kind was equivalent to the rank of *species* on the hierarchial level. However, as his knowledge of the living world increased, he felt that the species level was not as fixed as he had at first believed. Hence, in the final edition of Linnaeus' work, we find that he later changed his mind and now stated that in his present opinion, the *genus* was the hierarchial equivalent of the created kind. Regrettably, the scientists of that day did not agree with his revised opinion. As a result, it was Linnaeus' original position that persisted in the world of science. This view was also adopted by the church of that day.

The 19th Century

When we survey the immediate precursors of Darwin, the first we encounter is his grandfather, Erasmus Darwin (A.D. 1731-1802). He produced a two volume work entitled *Zoonamia.* It is interesting to note that in this work, he used the word 'evolution', although oddly enough, later Charles Darwin himself never used the word in the first editon of his *Origin of Species.* It is also interesting enough, to find that there are those who believe that Erasmus was really the first writer who proposed many of the main points of his grandson's theory.

Another immediate predecessor of Charles Darwin was Jean Lamarck (1744-1829 A.D.). He concluded that evolution is a general process that embraces every form of life. His theory was expressed in his *Histoire Naturelle des Animaux Sans Vertebres, 1815* by four laws:

1) Life by its proper forces tends continually to increase the volume of every body processing it, and to enlarge its parts, up to a limit which it brings about.

2) The production of a new organ in an animal body results from the supervention of a new want (besoin) continuing to make itself felt, and a new movement which this want gives birth to and encourages.

3) The development of organs and their force of action are constantly in ratio to the employment of these organs.

4) All that has been acquired, laid down, or changed in the organization of individuals in the course of their life is conserved by generation and transmitted to the new individuals which proceed from

those which have undergone these changes.

Lamarck's third law is more commonly expressed as the *Law of Use and Disuse,* in which he held that frequent and constant use of any organ will gradually strengthen, enlarge and thus develop that organ. This then will give it a strength proportional to the length of time of such use. On the other hand, lack of use of an organ will gradually weaken it, causing it to become reduced in size and function, and thus ultimately to disappear.

However, the main point of his theory was expressed in the fourth law, namely the *Law of Inheritance of Acquired Characters.* In this he further held that the advantages or disadvantages gained by any organism as the result of such structural changes would be handed down generation after generation. Obviously if this were true, there would also be a cumulative effect operating leading to the production of new forms.

Unfortunately, the social structure of the time was such that a great gap existed between the Chevalier de Lamarck and those breeders who would have had firsthand acquaintance with the progeny of animals. As a result there was little direct communication between these individuals and Lamarck during the time he was producing his 'Laws'. Hence plant and animal breeders soon were reporting that whatever might have happened to an animal in its lifetime would not affect the next generation. It is puzzling that Lamarck seemed unaware of the fact that various peoples had practiced mutilations of one kind or another for several millenia, for example, the circumcision of Jewish male infants, or the binding of the feet of Chinese girl babies. Yet despite such consistent long-time mutilations, there was never a record of any of these deformities appearing naturally in any child after a long period of such practices. This information was certainly known at the time. Later, August Weismann (A.D. 1834-1914) mutilated the tails of generation after generation of rats. Finally, after following through 20 generations, he quit in disgust, after finding that the tails of the last were as long as those of the first. Thus he demonstrated that any changes in the body, other than genetic, cannot be inherited. Nevertheless, Lamarck's theory was the only serious competitor to Darwinism for almost a century. He was much on Darwin's mind, and Darwin referred to Lamarck and his theory again and again. Actually, in the last writings of Darwin, his pangenes theory was much like Lamarck's, despite all of Darwin's previous contemptuous references to Lamarckism.

A contemporary as well as countryman of Lamarck's was Georges Cuvier (1769-1832). This brilliant French naturalist became known as 'the father of comparative anatomy' as well as 'the father of paleontology' because of his pioneering careful and able work in both of those fields of science. Cuvier was the first to compare systematically the structure of existing animals with the fossil remains of extinct animals. Therefore he came to the conclusion that if an incomplete fossil were found which was equipped with fangs, he could safely predict that all other parts, when found, would be consistent with those of a carnivore. In almost all instances, when the remaining parts of a fossil carnivore were found, it did turn out to have consistent parts, and Cuvier was proven correct. As he further studied fossils he became convinced that there was no evidence to support any evolutionary development among living organisms. Therefore he remained a creationist the whole of his active life. Lamarck's publication of his own theories led to a series of confrontations between himself and Cuvier. However, Cuvier's great reputation and known competence in the field of biology as well as paleontology, enabled him to prevail against Lamarck's evolutionism in favor of his own creationism.

Herbert Spencer (1820-1903), a contemporary of Darwin, was actually a philosopher rather than a scientist. But he also wrote on the subject of origins. He proposed a full-grown concrete theory of evolution, but without advancing any supporting scientific evidence. Hence, although his writings on the subject just preceded the *Origin,* the naturalists of his day would not accept his theory. Spencer and other philosophers of that day are credited with the first use of such phrases as 'survival of the fittest' and 'nature red in tooth and claw'.

Finally, one must certainly mention a British clergyman and naturalist named William Paley (1743-1805). He embodied his ideas in a well-known work entitled *Natural Theology* which was published in 1802. The whole of Paley's philosophy was based on the principle that just as the existence of a watch implied the existence of the watchmaker, so the human eye would imply the existence of a creator. He saw design in nature wherever he looked. In retrospect, his work is considered to have delayed for a considerable length of time the acceptance of any evolutionary theory prior to Darwin's day. Interestingly enough, Darwin was quite familiar with Paley's work. In his college days, he admired Paley greatly, so much so, that a copy of his *Natural Theology* accompanied Darwin on his voyage on the *H.M.S. Beagle.* Darwin's own statements indicated that throughout his working life, many of Paley's arguments haunted him[5].

---

[5]Charles Darwin, *Origin of Species*, Everyman's Library, Dutton, NYC, 1971, p. 167.

II. THE ADVENT OF THE *ORIGIN*

Charles Darwin

There are many different opinions about the work of Darwin (1809-1882), and many volumes of writings dealing with his life and work have been produced. His voyage around the world on the *Beagle* alone has been the subject of several works, and there is his own biographical account of the voyage. It was this experience that changed Darwin's thinking about the world of living things. While at college, he had believed what he was taught at that time, namely, that all living animals were found where they had been created, because in each case, God had created each animal in that location, and they had remained there ever since. Obviously this belief was not in harmony with what Darwin was observing on the voyage. Hence, although he began the voyage as a creationist, he arrived back in England with a greatly altered outlook on the living world. In retrospect, it is amazing that neither Darwin nor his teachers apparently read further than Genesis 1-3. Had they continued to the story of the Deluge, they would have realized that according to that account, obviously all air breathers perished in the waters, and each basic kind had to spread out from wherever the Ark landed.

This point emphasizes that evolutionists and creationists can share an interest in studying the problems of the geographical distribution of plants and animals, since both would agree as to the fact of such distribution. The creationist might work from the dispersion of animal forms after the Deluge, while the evolutionist would work from his point of view that animals would spread out from wherever any particular form had evolved.

The *Origin*

Darwin did not produce his great work, *The Origin of Species* until he was 50 years of age. This is one of the factors that relegates the well-known tract, *Darwin's Last Hours*, to the realm of wishful thinking. This tract quotes Darwin as saying that 'he had produced his work when an immature young man who didn't know any better.'[6]

A study of the main points of Darwin's theory will show that they can be summarized in the form of three observations and two deductions as follows:
    a) all organisms produce far more offspring then are able to survive, (Observation #1) and

---

[6]Wilbert Rusch, "Darwin's Last Hours," *Creation Research Society Quarterly*, XII, 2, pp. 99-102.

b) there are some of these that in turn will survive and produce offspring, while others will die, (Observation #2) and
c) those that survive do so because of small variations that are advantageous (Deduction #1); and further
d) those that have these advantages, by virtue of survival, can pass them on to their descendants, (Observation #3) and finally,
e) natural selection will sift out the variations that are favorable to survival and the sum of these will give rise to a new species. (Deduction #2)

The preceding outline is that explanation of evolutionism that became known as Darwinism.

One of the problems that develops in dealing with the theory of evolution is that it has taken on so many forms. One form of evolutionism became known as Lamarckism, as outlined by Lamarck. This in turn was superseded by Darwinism. Then it was realized that any innovations, due to natural selection, would inevitably be swamped out in future generations. This realization brought on a resurgence of Lamarckism in the form of Neo-Lamarckism. This was especially supported by the paleontologists. However, not being able to survive a revival of Weismann's observations, this form in turn was followed by the Synthetic Theory which appeared in the 1930's. This theory was a blend of natural selection and de Vries' mutation theory, with support from the finds of paleontology. Actually, the real proof of the theory of evolution must lie in the fossil evidence. But we find an increasing number of admissions regarding the inadequacy of this evidence ranging from T. H. Huxley, through G. G. Simpson to Kitts. Let us see what a relatively recent statement holds on the matter. We find David Kitts saying:

> "Despite the bright promise that paleontology provides a means of 'seeing' evolution, it has presented some nasty difficulties for evolutionists, the most notorious of which is the presence of 'gaps' in the fossil record. Evolution requires intermediate forms between species and paleontology does not supply them."[7]

As a result, today we find that cladistics and/or punctuated equilibrium of S. Gould and N. Eldridge are now the vogue. In the interests of clarity, it becomes necessary to pause at this point and

---

David B. Kitts, "Paleontology and Evolutionary Theory," *Evolution*, XXVIII, September 1974, p. 467.

recognize that what really are referred to thus far, are the theories proposed to *explain how* an evolution supposedly came about. Evolutionism at this point might be summarized as follows:

It is a theory which holds that;
    a) living matter spontaneously came from the non-living,
    b) living matter thus formed, changed through long eons of time, developing from a simple form to many complex forms,
    c) these complex forms are represented by all living forms as well as all fossil forms, and finally
    d) no one knows how this all happened, since none of the proposed explanations to date has held up.

It seems awfully hard for some to admit that maybe the reason that *none* of these various explanations have held up with the passage of time is because **EVOLUTION NEVER OCCURRED!**

III. THE AFTERMATH

The Acceptance

It is amazing that within a few decades after the appearance of the *ORIGIN*, Darwin's propositions were widely accepted. There might be several reasons for this happening. Thomas Huxley (1825-1895), who styled himself Darwin's bulldog, pointed out one. He had been a severe critic of several previous theories which had been proposed by various individuals, e.g. Robert Chamber's VESTIGES OF NATURAL HISTORY OF CREATION. Referring to this time, Huxley once wrote:

> "We wanted not to pin our faith to that or any other speculation, but to get hold of clear and definite conceptions, which could be brought face to face with facts and have their validity tested. The *Origin* provided us with the working hypothesis we sought. Moreover it did us the immense service of freeing us forever from the dilemma — refuse to accept the Creation hypothesis and what do you have to propose that can be accepted by any cautious reasoner?"[8]

Unfortunately for Huxley, the great promise of Darwin's theory that he had looked for, did not materialize through the years. Formidable objections to it arose as time went on, so that Huxley's point about creation really continued to remain valid.

From another point of view, George Bernard Shaw once wrote an opinion as to why Darwin's theory was so readily accepted. He stated

> "if you can realize how insufferably the world was oppressed by the notion that everything that happened was an arbitrary personal act of an arbitrary personal God of dangerous, jealous and cruel personal character, you will understand how the world jumped at Darwin'"[9]

What Shaw is pointing out was, that if there was a God who has created all things, it follows that He must be a very powerful Being, who can and does lay down His rules. Mankind must then obey or suffer under His justice. Now Darwin was demonstrating, through his theory, that God was unnecessary, that He had been replaced by nature and chance. What a relief to mankind. Man could now do as he pleased without fear of any reprisal from a God for his offenses against God. In

---

[8] Leonard Huxley, *Life and Letters of Thomas Henry Huxley*, Macmillan, NYC, Vol. 1, 1903, p. 180.

[9] Douglas Dewar and H. S. Shelton, *Is Evolution Proved?*, Hollis and Carter, London, 1947, p. 4.

fact there really was no more sin, since there was no one to make any rules. (See *Humanist Manifesto II,* found as Appendix D.)

The Effect on Religion

The acceptance of Darwinism had a direct effect on religion generally. If everything on earth was the effect of an evolution by chance, then man's various religions would have to develop under the same principle. Obviously acceptance of this thought meant that the authority of the Bible was undermined. It now became just another book, with no more authority than Homer's *Iliad* or any other classical works. As such, it was also subject to human evaluation and judgments. Thus the way was opened to secular humanism, which decreed that man was now master of his destiny, actually capable of directing his own evolution. There was neither God nor Ten Commandments to stand in the way. Man was now free to do as he pleased without the fear of any accounting.

In this connection a quote from Aldous Huxley, grandson of Thomas H. Huxley, might be pertinent. In 1937 he wrote in *Ends and Means* the following words:

> "I had motives for not wanting the world to have a meaning; consequently assumed that it had none, and was able without any difficulty to find satisfying reasons for this assumption.... The philosopher who finds no meaning in the world is not concerned exclusively with a problem in pure metaphysics; he is also concerned to prove that there is no valid reason why he personally should not do as he wants to do ... for myself, as no doubt for most of my contemporaries, the philosophy of meaningless was essentially an instrument of liberation. The liberation we desired was simultaneously liberation from a certain political and economic system and liberation from a certain system of morality. We objected to the morality, because it interfered with our sexual freedom; we objected to the political and economic system because it was unjust. The supporters of these systems claimed that in some way they embodied the meaning (a Christian meaning, they insisted) of the world. There was one admirably simple method of confuting these people and at the same time justifying ourselves in our political and erotic revolt; we would deny that the world had any meaning whatsoever."[10]

As a result of the widespread adoption of this materialistic

---

[10] Aldous Huxley, "Ends and Means," quoted from R. F. R. Gardner, *Abortion, The Personal Dilemma,* Eerdmans, Grand Rapids, 1972, p. 57.

philosophy, so often we hear in our country today this modern view expressed in the words 'if it feels good to you, if you like it, if it harms no one, do it!' (Again, see *Humanist Manifesto II,* Appendix D.)

Jacques Barzun in his *Darwin, Marx, Wagner* showed that the philosophies of Marx and Nietzsche were natural outgrowths of Darwinism. It is also interesting to note that when Karl Marx read the *Origin,* he was wildly enthusiastic, regarding it as bolstering his own theories. In fact, he wanted to dedicate his work *Das Kapital* to Darwin, but Darwin politely declined the honor. Although evolutionism was widely accepted in Great Britain, Germany and the United States, it is a little realized fact by those in the English speaking countries, that France seemed to retain a skeptical position on the whole matter. This is still true today.

The French equivalent of the Encyclopedia Britannica, was published by the Paris firm of Larousse under the title *ENCYCLOPEDIE FRANCAISE.* The editor of the 5th volume entitled *Plants and Animals* was Paul Lemoine. In his concluding essay *What Are the Theories of Evolution Worth?* he had this to say:

"The theories of evolution with which our studious youth have been deceived, constitute a dogma that all the world continues to teach, but each in his specialty, the zoologist or the botanist ascertain that none of the explanations furnished are adequate."[11]

In his conclusion in the final chapter he states:

"It results from this summary that the theory of evolution is impossible".

It should be stated, that Paul Lemoine is still speaking as an evolutionist who nevertheless still has faith in evolutionism. Although having made the critical statements cited, he does not thereupon espouse creation.

The Scopes Trial

In the United States, some states responded to the acceptance of the theory of evolution by passing legislation, forbidding the teaching of evolutionism in the public schools. In 1925, this situation led to the farce known as 'The Scopes Trial'.[12] This trial has been re-examined recently in numerous journals. The outcome of the trial was that Scopes was declared guilty of teaching evolutionism. However, in appearances

---

[11] *Encyclopedie Francaise,* Larousse, Paris, V, pp. 82-83.
[12] R.M. Cornelius, "Their Stage Drew All the World," *Tennessee Historical Quarterly,* XL, Summer 1981, pp. 1-15.

on TV talk shows shortly before his death, he had admitted that he feared that he might be called to the stand to testify, in which case he would have had to have stated that he had not actually taught the subject. Fortunately for him, no one thought of asking him that question. In fact, the evidence in the recent studies show that the whole affair actually was a put up job by enemies of the Scriptures, who wanted to see them discredited. As far as the achievement of this aim is concerned, in many circles, the trial was a success.

Oddly enough the trial affected the teaching of biology on the high school level in the public schools very little. A perusal of high school biology texts, both before and after the trial, will reveal that during both periods, the following situation prevailed; in some texts the subject of evolutionism was never mentioned, while in others the subject was restricted to a chapter or two. In the second instance, the subject could be taught or not as the teacher was inclined and as the spirit of the area where the schools were located dictated. This sort of armed truce continued into the early 1960's.

The Darwin Centennial

The year 1959 was the hundredth anniversary of the publication of Darwin's *Origin of Species*. Therefore this year was designated the Darwin Centennial Year. The headquarters of the celebration were on the campus of the University of Chicago, under the direction of Sol Tax, anthropologist in residence. During this time various programs were arranged to do honor to Darwin by means of lectures, seminars, panel discussions, etc. The complete record of the events is available in a three volume work edited by Sol Tax.[13]

One of the celebrations for the Centennial was the AAAS (American Association for the Advancement of Science) banquet, with Dr. H. J. Muller, (1890-1967) noted geneticist at Indiana University, as the principal speaker. His topic was *One Hundred Years Without Darwinism Is Enough!* In his address Muller bewailed the fact that he found students in his undergraduate classes in genetics who were not only in many cases ignorant of the principles of evolution and didn't believe it, but many of them had never even heard of it! He considered this an outrage, hence his topic. He then sounded a clarion call to the assembled members and guests to rectify this state of affairs as soon as possible, regarding this action as long overdue.

The BSCS and Its Work

Essentially as a response to Muller's call for action, a group was

---

[13]*Evolution After Darwin*, Sol Tax, Ed., Un. of Chicago Press, Chicago, 1960, 3 vol.

organized in Boulder, Colorado, which was called the BSCS (Biological Science Curriculum Studies). The upshot of their work was the production of three new biology texts, which were to rectify Muller's problem. They appeared as:
1) the *Blue Version,* with the heaviest emphasis on molecular structure and biochemistry;
2) the *Yellow Version,* with an emphasis on ecology, not an evil end in itself; and
3) the *Green Version,* with a more balanced topic distribution than either of the other two.

These were produced with extensive federal funding. From a strict pedagogical point of view, these texts were an improvement on the methodology of the old texts, especially in the area of laboratory exercises. However, the changes also involved the elimination of a good deal of the natural history content of biology with the substitution of considerable space devoted to the discussion of highly technical molecular structures with the attendant biochemical emphasis. It is open to question as to whether the elimination of much of the earlier natural history content was really desirable. In respect to creation, the approach used in these texts was a disaster. The texts assumed the theory of evolution as being utterly factual, and arranged the sequence of subjects according to the order of evolutionary development. None of the glaring weaknesses of the theory were ever presented. Especially in the case of the 'Blue' version, a creationist would find it almost impossible to use these texts for teaching biology. Obviously this situation would require the private Christian schools to develop their own texts, and their lack of financial resources in most instances prevented them from taking this recourse.

In terms of the public school, the situation changed from what might have been regarded as an armed truce to an outright declaration of war. One can imagine the feelings of outrage felt by many parents when their children began to come home to tell them that they were now taught that the creation story was utterly unscientific, as well as being impossible, and was little more than a fairy tale. Further, that instead they came to be on earth through a series of accidents, that when proper atmospheric conditions and the right chemicals accidentally came together, life was produced. Further, through a series of genetic mistakes, life became an ever increasing complex series of forms, and this is how they, the children, came to be on earth. All of this being presented as if it were utterly factual.

Outraged parents in California banded together and pressured school boards, legislatures, and the like for the safeguarding and protection of their and their children's rights. The main attack was against setting up a theory as a fact, over against a belief in creation. In many instances parents were successful in their quests, and some laws were passed safeguarding the rights of children to be free of ridicule for their beliefs, as well as compulsion in their choice of theories of origins. What has followed since, of course, has been a series of legal battles, which are still continuing in various parts of the country. The most recent instance was the trial in 1981 in Little Rock, Arkansas. The issue was the constitutionality of the Arkansas creation-evolution Act 590. The ACLU challenged the act and following the trial, the judge decided that the act was unconstitutional. Norman Geisler has written one of the few accurate accounts of this trial.[14]

The BSCS texts are no longer known by their colors. In fact the organization which originally handled their publication apparently was not too careful with their manner of disbursing funds, and hence came under question. The end result of the whole affair was that the publishing rights were given over to private publishing companies. Since naturally the publishers were primarily concerned with sales, changes were made. Some of the emphases on subject matter were altered. For example, there was an emphasis which was too heavy on biochemistry. This resulted in the unfortunate circumstance that numerous prospective young biologists, primarily interested in the natural history aspects of biology, were turned off by this approach and a number of promising biologists were lost to work in this field. Interestingly enough, in the several current textbooks that the writer has examined, no clue is given as to which if any of the original BSCS texts they were based on. Many such have obviously been greatly modified and toned down since their first advent.

---

[14]Norman Geisler, *The Creator in the Classroom*, Mott, Milford, Mich., 1982.

IV. TERMS AND ASSUMPTIONS

Creation-Science and Evolution-Science

It will be noted that the terms *creation-science* and *evolution-science* are not used by the writer in this paper, since he has an aversion to the use of such terms, and questions their validity. It seems that such usage would be analagous to using the terms *big bang astronomy vs. steady state astronomy;* or in a previous time, using *Copernician science* vs. *Brahian science.* In most common usage there is a discipline called *science,* and no particular paradigm or school of thought is conventionally recognized as another branch of that science. In the opinion of this writer, both evolutionists and creationists may study, experiment, and theorize about various aspects of the field of science. However, there is neither need nor justification for any unique terminology such as *creation-science* or *evolution-science.*

True Creation-Science?

The writer often has been intrigued by the possible validity of the designation of all of the science of pre-1870 as *scientific creationism.* Not that he is by any means seriously recommending this, but it often is overlooked by many evolutionists that that terminology well could be valid within the rules of science. After all, it is forgotten that creation was the ruling paradigm prior to the advent of Darwinism. To speculate further along these lines, the writer notes that there is a rule in science that says that if the existence of a prior meaning of a certain concept can be demonstrated, then according to the *law of priority,* the earlier meaning is the valid one for that term or concept.

For example, in his earlier days of teaching biology, the writer, followed the usage at that time of considering the term *symbiosis* as meaning simply the opposite of parasitism. Then it later was determined that the person who originally defined the term, meant that the term *symbiosis* was to have a much wider application. Actually it was to include all of the possible interrelationships among various organisms. Therefore the term *symbiosis* would include the concepts of mutualism, commensalism, and parasitism. Since this wider meaning was the prior one, that meaning then became the valid one. As a result, all of those who taught biology at that time had to adjust to the new definition. Another example of such changes in meaning would be the several changes in the definition of osmosis for similar reasons.

Continuing this thought, one could look at Francis Bacon, the father of modern science, who in referring to the study of science, had this to say:

"To conclude, let no one weakly imagine that man can search too far, or be too well studied in the book of God's word and works, divinity or philosophy; but rather let him endeavor an endless progression in both, only applying all to charity and not to pride."[15]

We must remember, that what we call science today, in those days was referred to as natural philosophy.

The Law of Priority

Furthermore, if one studies the works of Isaac Newton, one finds ample evidence that this was the general approach to science in those days. As his writings reveal, Isaac Newton believed that he was thinking the Creator's thoughts after Him, as he quested after scientific knowledge. In fact, that was his purpose in studying science! Newton was not alone in holding this point of view. This was a common line of scientific thought in those days.

Now if this can be substantiated as being the concept of science at that early time, then we might have an interesting application of the *Law of Priority*. Under the operation of that law, a person could justly say that after looking at what passes for science today, that it certainly isn't science according to the original concept of Bacon and Newton and their contemporaries. Therefore under the law of priority, there would be the requirement that a new name be found for what passes as science today, and according to priority, only scientific creationism would be *Science*!

This early state of affairs is frequently forgotten entirely or else deliberately ignored by most macroevolutionists. Especially is this demonstrated when so many today cast ridicule on the very thought of any creationists being considered as competent scientists.

Yet certainly the work of such creationists as Louis Pasteur (a pioneer in microbiology), Georges Cuvier (the father of paleontology), and Louis Agassiz (the father of glaciology), just to mention a few, would demonstrate the fallaciousness of such expressions. In addition, such men as Newton, Lister, Faraday, Mendel, Maxwell, Fleming, etc., are all well-known scientists who were pioneers in their various fields. Would macroevolutionists then presume to regard these men as unworthy of being called scientists since they were creationists? Further, if we remember that various pioneers in electricity, such as Maxwell, Faraday, and the like were all creationists, we would have achieved all the progress and benefits in this area without the contribution of any

---

[15]Francis Bacon, *Advancement of Learning*, Colonial Press, NYC, 1900, p. 5.

evolutionary thought whatsoever. Still again, certainly no one seriously will contend that our present-day computer explosion came about only because of the prevalence of the evolutionary paradigm.

Early Views of Science

It should be remembered that the subject matter, in the various fields mentioned in the previous paragraphs, fits the classical concept of science. This held that science was based on observation and experiment. Such an early basic concept of science was first described by Francis Bacon in his *Novum Organum*. He considered also that the purpose of science included the searching for ultimate causes, i.e. what they called the mathematical harmonies in the mind of the Creator. In the philosophy of such men as Bacon and Newton, science was not divorced from theology. Actually, the latter was often referred to as the *queen of the sciences*. Scholars of that day seemed to feel that it was necessary to take the whole aspect of life (man and *all* aspects of his environment) which also included the spiritual. It might be well to recall that some writers on the history of science, among them Whitehead[16] and Oppenheimer[17] have stated that the development of modern science would not have been possible without the climate of Christianity, with its concept of a God of order. They have pointed out that the world of the ancients, populated by the gods of mythology, was a world where the possibility of formulating natural laws would have been impossible. As long as man believed that supernatural beings, subject to the same passions that ruled human beings, governed the earth in a capricious and often vengeful manner, the organized study of natural phenomena with the inclusion of laws would have been impossible.

Science

Like so many words we use today, it is not certain that the precise meaning of the word "science" is always understood. The writer was once a member of a graduate seminar in geology, where the group's first day's assignment was to arrive at a definition of the term science. There were some 26 members in the seminar, consisting of both graduate students and faculty members. Yet the next meeting produced some 26 different opinions of the meaning of science. Following a good deal of discussion, the group narrowed the field down to two candidates for the definition. They were as follows;

One group held that science is the total *body of knowledge* of

---

[16] Alfred N. Whitehead, *Science and the Modern World,* Macmillan, NYC, 1926, p. 18.
[17] J. Robert Oppenheimer, "Science and Culture," *Encounter, Oct. 1946, p. 62.*

man's environment, i.e. the earth, life on it, and the space surrounding it.

The other group maintained that science is really a *method of investigation,* following almost the traditional steps of the scientific method, much as first laid out by Francis Bacon in his *Novum Organum.*

Today, the scientific method is considered as being composed of the following steps;
  a) the recognition of some problem or phenomenon that needs explaining,
  b) the gathering of all available information or data pertaining to the problem or phenomenon,
  c) the development of a propositional hypothesis from such data to explain the problem,
  d) and the planning and execution of a series of experiments to test the hypothesis.

This hypothesis is then re-examined and revised as called for by the experimental results. (NOTE: The hypothesis is revised, not the experimental results!) After a period of time, during which testing by sufficient experiments has been performed and checked for the validity of the hypothesis, it may be elevated to the level of a theory. It is held today that any hypothesis must have predictive value, which then can form the basis for further experimentation. An integral part of the experimental procedure is that any experiment, both the original as well as the followup, must be *repeatable,* i.e. they must be capable of being performed by others under the same conditions with the same results. Only then can such results be considered valid. It is obvious at this point that the above procedures are not applicable in the case of much of the study of origins. Therefore in such matters, the last word never really can be said. It actually is questionable whether any of the study of origins is truly science. Therefore this whole matter would seem to belong to the area of philosophy to a great extent.

We could combine the two definitions previously expressed, by defining *science* as implying the collective human knowledge in any field of study. However, the term is ordinarily applied to any organized field of study, which is investigated by the scientific method, as outlined previously. It also would include any practical application of the body of facts obtained by such investigation. Since we are forced into an increasing specialization of endeavor in our day, science is divided into a large number of fields, and each of which current usage terms a *science.*

Two Aspects of Scientific Inquiry

A greater understanding of the difficulties in semantics previously referred to develops when it is realized that there are really two aspects to scientific inquiry. One aspect deals with the here and now. The other aspect is encountered in the large areas that to a greater or lesser degree involve the past. Scientific inquiry into the here and now concerns problems which can be studied, observed, measured (e.g., as in natural history, where life histories of plants and animals are studied; also identification of rocks and minerals). Such problems also can be the objects of experiment, and in some fields, treated mathematically (e.g., as in physics and chemistry). In other words, in these areas one uses the scientific method. With regard to the actual data, there need be no disagreement between creationists and evolutionists. The subject matter is factual and subject to objective observation. As far as assumptions are concerned, we assume that what is seen must be believed. In these areas, other assumptions are few and far between.

As a case in point, the writer can recall that many years ago, he was attending a state university as a graduate student majoring in biology. In the majority of the courses taken, the question of evolutionism was never even raised. It was possible for a creationist to work profitably side by side with evolutionists in apparent harmony. The writer is sure that this was not an unusual circumstance.

Speculative Scientific Inquiry

On the other hand, examples of that aspect of scientific inquiry where large areas are dealt with in an attempt to understand the past would include: much of historical geology, a good deal of structural geology, paleontology, astronomy (other than descriptive), etc. In studying the past, one cannot use the scientific method, because one cannot observe past phenomena as a basis for drawing conclusions. One can use only one's theories of past phenomena, (which may or not be true) as a basis for one's conclusions. This process is hardly the scientific procedure as suggested by Bacon. By no stretch of the imagination can this sort of exercise be classed as empirical science. One could only classify such fields as belonging in science in the broad sense, distinguished from science in the strict or narrow sense.

If one were to ask a student of paleontology studying the fossils he might have found in a rock, what he was studying, his answer would likely be "the life of the past." However he is really not doing anything of the kind. All that he actually is studying is *the peculiar items imbedded in the matrix of the rock,* no more. On the basis of his observations, he could draw conclusions as to what those inclusions might be, whether or not they were once living, and what the conditions were under which they lived, etc. The answers to these questions are in the area of conjecture, and of necessity, would remain there. This is why the history of paleontology is replete with examples of conclusions drawn, which further evidence proved untenable. This also applies to reconstructions which were made and then later rescinded as further evidence appeared. It must be said that in modern times, this sort of occurrence has rarely occurred other then in the case of pre-Cambrian fossils. As one example, consider the history of the discovery of pre-Cambrian fossils. Here many fossils, supposedly living at that time, were hailed as representing the earliest forms of life. Later corrections had to be announced, when the supposed fossils were identified as non-living inclusions.

One of the earliest of these supposed Pre-Cambrian fossils to be described was named *Eozoon canadense* by its discoverer, Dawson. He regarded it as a giant foraminifer. For years after, this fossil was generally accepted as a genuine Pre-Cambrian find by numerous geologists, Darwin among others. However, over 40 years ago, it was rejected. Already in 1935, Dr. Percy Raymond, in his presidential address stated to the Geological Society of America:

> "It is obvious that *Eozoon* is the product of two periods of alteration of the original sediment and can by no possibility represent an organic structure. There seems not the slightest chance that it can be organic."[18]

The most recent example is the fate of the supposed fossilized yeast-like cells found in the Isua belt in Greenland. These were identified as yeast cells by H. D. Pflug, and he named them *Isuasphaera isua*. Examination of the photographs of the structures showed that they looked much like living cells.[19] Since the Isua belt is considered the earth's oldest rock outcrop, Pflug believed he had found the oldest life.

Unfortunately, for this proof of early life, B. Nagy and his geochemistry colleagues at the University of Arizona identified dozens of amino acids found in pulverized Isua rocks. The types of amino acids found were all the L variety, thus indicating that Pflug's supposed microorganisms were only a few tens of thousands of years old. Also, in all probability, they were not even a life form, since laboratory examination of the rocks indicated that the organic compounds identified could not have survived the known metamorphic history of the Isua rocks. In

---
[18] Percy Raymond, *Bull. Geol. Soc. of America*, Vol. 46, p. 378.
[19] H. D. Pflug & H. Jaeschke-Boyer, "Combined Structural and Chemical Analysis of 3,800 Myr. Old Microfossils," *Nature*, Vol. 280, pp. 483-6.

fact the Nagy group considered that the amino acids were the result of diffusion into the rocks from surface lichen.[20]

It should be emphasized that the reconstructions of fossil vertebrates which are so often seen in museums, popular journals, and texts should be taken with a large grain of salt. Since a characteristic of vertebrates is bilateral symmetry, if half of the fossil skeleton is found, it is legitimate to reconstruct roughly the size and form of the animal. But such details as muscle distribution and form, skin texture, and color can simply not be known. In complete reconstructions they are probably the product of one authority's imagination. For example, it is rarely realized that only one part of a single dinosaur's skin is known. In this particular instance, the scales were found to be somewhat beaded like those of a Gila monster. What the skins of all of the other dinosaurs were like, is not known. The same is true of the mammals found as skeletons.

Obviously, phenomena that have occurred in the past cannot be observed directly, nor can they be the subject of experiments. Therefore, the result of the study of any evidence of past phenomena yields only a series of deductions made up of inferences, supposedly reconstructing circumstances, which in turn rest on assumptions, any of which may or may not be true. The writer has no objection whatsoever to this practice; in fact he can see the necessity and desirability of it to a great extent. But the problem comes with the subsequent treatment of the whole result, particularly as it presently appears in texts at the elementary and secondary levels. Here we often find personal convictions, remote possibilities, wild speculations, all of them presented as if they were factual proofs, or at least valid arguments in favor of evolutionism. Thus many speculations of that nature have been perpetuated through the years, built up to veritable mountains of hypotheses that are thought of and taught as being factual.

One has only to think of straight examples from older texts as the many illustrations of straight-line horse evolution, the recapitulation illustrations, the phylogenies of various kinds, much of the fabric of historical geology, etc. The writer is not suggesting the elimination of the whole of this general area in science, rather he is recommending the retention of a proper perspective and sense of values through the whole, with the elimination of the appearance of actuality where none exists. However, we have had outright fraud as in Haeckel's 'Biogenetic Law' with its illustrations of developing embryo series. Haeckel actually had the same woodcut printed three times to illustrate the similarity between

---

[20]Bartholomew Nagy, et al., "Amino Acids and the Hydrocarbons in 3,800 Myr. Old Rocks in the Isua Rocks, Southwestern Greenland," *Nature*, Vol. 289, 1/8 Jun. 1981, pp. 53-55.

embryos of chicken, tortoise, and dog. The embryo series Haeckel produced has been condemned as false, yet in some form or another they still continue to appear in lower level texts.[21] It is outrageous that these illustrations should still be perpetuated. When speculations of such a nature are presented in a lower-level text, they should be clearly identified as being only someone's tentative suggestions, subject to change without notice. A perusal of the texts on these levels reveals a general absence of such honesty.

Uniformitarianism

An underlying principle in the development of evolutionism has been that tremendous assumption, known as *uniformitarianism*. Succinctly it is the idea that *the present is the key to the past*. Having frankly first been considered an assumption, it later often was given the status of a law. Actually, the present *may or may not* be the key to the past. In the professional journals of science it now is recognized that the uniformitarianism concept is utterly dead, having been superseded by the principle of *uniformity,* which is something quite different. Proponents of uniformity grant that the rates at which phenomena may have occurred could have been drastically different in the past. They even allow for some catastrophism. However this new concept has limits as to how far it can be pushed. Obviously in any field of such speculative inquiry as dealing with the past, there is no guarantee of scientific precision or accuracy. Yet is is just in these areas that the question of origins arises. It then follows that nothing certain can be said about origins. All that is done is to produce one interpolation after another, each by its very nature being unverifiable. It would appear reasonable to propose that this whole area should not really be in the realm of science as strictly defined at all, but rather in the domain of philosophy (or science fiction)! As was stated previously, such 'sciences' are not empirical sciences. However, we must face the fact nevertheless, that these fields generally are considered legitimate parts of science as it is known today.

---

[21] Wilbert H. Rusch, "Ontogeny Recapitulates Phylogeny," *Creation Research Society Quarterly,* VI, 1, pp. 27ff.

V. EVOLUTIONISM

Definition of Macroevolution

The theory of macroevolution is difficult to define, essentially because the subject is so complex. If you were to pick a definition from one text, you would find a rather lengthy wordy statement:

> "Organic evolution is a series of partial or complete and irreversible transformations of the genetic composition of populations, based principally upon altered interactions with their environment. It consists chiefly of adaptive radiations into new environments, adjustments to environmental changes that take place in a particular habitat, and the origin of new ways for exploiting existing habitats. These adaptive changes occasionally give rise to greater complexity of developmental pattern, of physiological reactions, and of interactions between populations and their environment."[22]

Olson and Robinson make the astonishing statement in their listing of important concepts that:

> Evolutionism is the doctrine that the universe, including inorganic and organic matter in all its manifestations, is the product of gradual and progressive development.[23]

Some decades ago, Dobzhansky gave a rather concise and clear definition by stating that the theory embodied the following;

1) beings now living have descended from different beings which have lived in the past,
2) the discontinuous variations observed at our time level have arisen gradually,
3) all these changes have arisen from causes which now continue to be in operation, and which therefore can be studied experimentally.[24]

Popular journals, when they speak of the theory of evolution, consider it as embodying an 'ameba to man development'. However, beginning with Standen's discussion[25] in his *Science Is a Sacred Cow*, where he pointed out that there was a vague theory and a precise theory of evolution, we find that there has come to be a dual meaning to the

---

[22] Theodosius Dobzhansky, Francisco Ayala, G. Ledyard Stebbins, and James W. Valentine, *Evolution*, Freeman, San Francisco, 1977, p. 8.

[23] Everett C. Olson and Jane Robinson, *Concepts of Evolution*, Merrill, Columbus, 1975, p. 10.

[24] Theodosius Dobzhansky, *Genetics and the Origin of Species*, 1941, 2nd ed., p. 7-8.

[25] Anthony Standen, *Science is a Sacred Cow*, Dutton, NYC, 1950 p. 101ff.

word. This was best expressed by G. A. Kerkut, when he wrote as the concluding paragraph in his *Implications* the following:

"There is a theory which states that many living animals can be observed over the course of time to undergo changes so that new species are formed. This can be called the "Special Theory of Evolution" and can be demonstrated in certain cases by experiments. On the other hand there is the theory that all living forms in the world have arisen from a single source which itself came from an inorganic form. This theory can be called the "General Theory of Evolution" and the evidence that supports it is not sufficiently strong to allow us to consider it as anything more than a working hypothesis."[26]

Obviously Kerkut is recognizing that variation can be divided into two kinds:

a) one on a lower hierarchial level, (variety, species, genus and even family)
b) another at the higher hierarchial levels (phylum, class, order).

Today, one finds that almost all authorities agree with this concept; only the terminology has been changed. Today Kerkut's *Special Theory* is called *Microevolution*, and his *General Theory*, is called *Macroevolution*. Since the writer considers this new terminology the more precise and meaningful, throughout the remainder of this paper, the new terminology will be used. He further suggests that all creationists do likewise.

Early Arguments for Macroevolution

Until rather recently, texts would parade the list of arguments for macroevolution. Traditionally, these were classification, embryology, homology, genetics, mimicry, etc. But above all there was the fossil evidence. Yet, the early evolutionists, including Darwin himself, recognized the weakness of many of the various arguments they advanced favoring macroevolution. But they believed that these would be substantiated more and more as time went on, and as additional knowledge was gained. This is particularly true in Darwin's discussion of the fossil evidence which he produced in support of his theory. However, even in his day, he admitted the weakness of the evidence. He spoke several times of the incompleteness of the fossil evidence, which resulted in the lack of the transition forms that his theory called for.

---

[26]G. H. Kerkut, *Implications of Evolution,* Pergamon, NYC, 1960, p. 157.

However, not only Darwin, but all the Darwinists following were confident that time would bring the discovery of additional fossils, which in turn would bolster their arguments. However, the promised substantiation of the various evolutionary phylogenies by fossil finds simply has not materialized. Many other arguments bolstering the theory have fallen by the wayside. Norman Macbeth has amply demonstrated that in the minds of most biologists today, classical Darwinism is dead.[27]

But what about the successor to Darwinism, namely the Modern Synthesis? A meeting was held at Chicago in October, 1980 with the simple conference title, *Macroevolution*. Here some 150 paleontologists, anatomists, taxonomists, embryologists and geneticists debated as to whether or not macroevolution could be explained as the mere extension of microevolutionary changes. I doubt whether Kerkut's name was mentioned, but his statement of twenty years previous came to mind. This as "it is not clear whether the changes that bring about speciation (microevolution) are of the same nature as those that brought about the development of new phyla" (macroevolution). (Parenthetical statements are the writer's)[28]. The continuing pertinence of his statement was emphasized by the recorder of the Chicago meeting. His words are as follows:

> "The central question of the Chicago conference was whether the mechanisms underlying microevolution can be extrapolated to explain the phenomena of macroevolution. At the risk of doing violence to the positions of some of the people at the meeting, the answer can be given as a clear, No."[29]

However, let there be no misunderstanding, there was no doubt in the minds of most of those present that macroevolution had occurred.

---

[27]Norman Macbeth, *Darwin Retried*. Gambit, Boston, 1971, pp. 4, 5.
[28]G. H. Kerkut, 1960, p. 157.
[29]Roger Lewin, "Evolutionary Theory Under Fire," *Science*, Vol. 210, Nov. 21, 1980, p. 883.

VI. Creationism

Defining Creationism

At this point the writer will undertake a difficult task. Just as a perusal of numerous texts on evolution or biology will reveal a lack of agreement on what *evolution* is, so one will find the same situation among creationists. Yet, it is important to arrive at a general definition of creation.

Creationists probably would agree that the term includes these basic considerations. *Firstly, the concept of creation implies, of necessity, the existence of a Creator, an all-powerful Being.* This would be a *sine qua non* for the definition. It is well to remember that Paley operated on this principle, when he pointed out that just as the existence of a watch implies the existence of a watchmaker, so the existence of the universe implies the existence of a Creator.

Secondly, this Creator set into motion the basic laws according to which the universe operates and also sustains them. This Creator would be called by many names, be it the Christian's Triune God, the Unitarian's Supreme Being, the Hebrew's Yahweh, the Muslim's Allah, or the ancient Egyptian's Ammon-Re, i.e., 'he who made all', as well as others.

Thirdly, creation also implies that this Creator called into existence basic kinds of animals, which would include living as well as fossil forms. These basic *kinds* would be distinct, and would not grade into each other. The basic kind would not be equivalent to the biologist's species, but would transcend that concept (See pp.30-31). Therefore the term 'baramin' could well be applied to this category. This whole scenario would be diametrically opposed to that for macroevolution, since the latter implies a continuum, which creation does not.

In accordance with the basic ideas so far considered, we would expect to find exactly what is found as the fossil evidence, namely that organisms seem to appear in sequences of the rock strata suddenly, with no evidence in underlying strata of the previous forms from which they are said to have been derived. E.g. ammonoids appear suddenly, vary and then become extinct without ever having been anything other than ammonoids. There is nothing but speculation as to what invertebrate group gave rise to the vertebrates. These also appear in the strata suddenly and without previous ancestors. There are no fossil lineages that clearly delineate which of the invertebrate groups have been derived from which other invertebrate groups. It is not generally realized, but all the major invertebrate phyla were established during just one period, the Cambrian. This includes the arthropoda, made up of very complex organisms, such as the trilobites.

Another point is that in the fossil evidence, we find that the whale line is always distinct from every other line of mammalian development. In other words, there is no record as to which particular mammalian ancestor the whale evolved from. Again, the icthyosaurs are also always distinct from every other line of reptiles. Neither are there any intermediates between turtles and lizards, or between lizards and snakes, or between turtles and snakes. Yet these are all reptiles, and if macroevolution be true, the fossil evidence should show a continuity of development between them. But although these problems plague the evolutionist, they do not present a problem for the creationist, rather they are just what he expects to find in the fossil evidence.

Finally, a creationist would hold that the Creator incorporated into the basic kinds of organisms the ability to vary to a considerable extent, but only within the framework of the kind. Hence the variation according to creation would be limited, in contrast to macroevolution, which requires an unlimited variation for the theory to work. Creation holds that each basic kind was specially created. Living forms of various basic types are found to vary to a greater or lesser extent, and they may also survive, or become extinct. Such variation, however, is bound within limits, as is consistent with the finds of genetics and the experience of animal breeders. This limitation is also consistent with Goldschmidt's findings discussed elsewhere. Therefore it should be obvious that the creationist would find little to quarrel with in the concept *microevolution*. But he would be in utter disagreement with *macroevolution*. The writer would prefer to refer to microevolution as 'variation within fixed limits' while macroevolution would be unlimited variation.

I believe that this general view of Creation would be common to scientists within the ranks of Christianity, Islam, Orthodox Judaism, and many other religious groups. There will be variations as to how creation came about, there will be variations in what names the various groups may bestow on the Creator, and there will be variations as to when they believe that creation took place. However, the four basic considerations mentioned previously form the basic kernel.

Is Creation Religious?

It is necessary to distinguish between Biblical creationism and a creationist discussion of empirical evidence; the latter is not religious and does not include a description of the Creator. Equally, not all definitions of religion entail the concept of a Supreme Being. The U.S. Supreme Court has this issue before it in 1960 in the matter of one, Torcaso. Here the court ruled that those who do not believe in God can still have a conscientious objector status on religious grounds. The record on page 987, paragraph 11, reads:

> "Among religions in this country which do not teach what would generally be considered a belief in the existence of God are Buddhism, Taoism, Ethical Culture, Secular Humanism and others."[30]

However, let us suppose for the sake of argument, that creationism is religious. That *per se* would have nothing to do with the validity of the creation paradigm as being scientific. As has been previously pointed out, science progressed admirably for centuries under the creation paradigm. In this connection it is rather amazing to find that Thomas Huxley once wrote:

> "... 'creation' in the ordinary sense of the word is perfectly conceivable. I find no difficulty in conceiving that, at some former period, this universe was not in existence; and that it made its appearance in six days (or instantaneously if that is preferred), in consequence of the volition of some pre-existing Being. Then, as now, the so-called *a priori* arguments against Theism, and given a Deity, against the possibility of creative acts, appeared to me to be devoid of reasonable foundation."[31]

This is certainly an astonishing, but frank statement. It is astonishing, especially when one considers that Huxley made it!

In any case, Darwin and his followers proposed that a Creator was no longer necessary for an explanation of origins. As has been outlined previously and will be pointed out as the paper moves along, the substantiation of the various 'proofs' of macroevolution simply have not materialized. It follows that macroevolution stands in no better state of certainty than creation as an explanation of origins, and therefore does not deserve the adulation that the scientists of today heap upon it.

Scientific Status of Creationism and Evolutionism

Initially, the point should be made that creationism and evolutionism, as explanations of origins, are *both* in the category of the speculative. Neither is substantiated on the level, for example, of the laws of classical physics. Yet one encounters a lot of arrant nonsense, written by those who ought to know better, to the effect that evolution is as well substantiated and factual as the law of gravity! At present, evolution is still only a statement of *belief* as to the origin of the various

---

[30] U. S. Supreme Court Reports, 367 US 488, 6 L ed 2d 982, p. 987.
[31] Leonard Huxley, 1903, Vol. 1, p. 241.

forms of life, both past and present. Some, looking at the natural environment, believe that Darwin's proposal is the only valid explanation for the origin of our living world. But it should be remembered that Darwin's explanation was based on both his numerous assumptions and his interpretations of various natural biological and geological phenomena. Therefore it would be more correct to say that some of the scientific evidence in this area supports creationism and evolutionism equally well, while some other data support evolutionism to a greater extent than creationism, and finally, certain items of evidence support creationism more than evolutionism.[32] It is the writer's opinion that the preponderance of evidence today would favor creationism. He will also agree that this is just his opinion.

---

[32]E. J. Corner, "Evolution," *Contemporary Botanical Thought*, McClead & Cobley, ed., Quadrangle, Chicago, 1961, p. 97.

VII. THE LIMITS OF VARIATION

Linnaeus' Original Position

Returning to the basic discussion of origins, one important point must be clarified before an intelligent discussion can be held on the two theories of origins. This is the definition of what might be called the basic unit of creation. It will be recalled that it was Linnaeus in his *Systema Naturae,* (1758) who first made a definitive statement on this matter. It was his considered opinion, as a result of the summation of all his observations up to that time, that the original created kind was equivalent to the rank of *species* on the hierarchial level, although in the final edition of Linnaeus' work, his *Systema Vegetabilium* (1774 edition) we find that he changed his mind. In this work, he held that what he referred to as the *natural order* was the hierarchical equivalent of the created kind. Regrettably however, the scientists of that day did not agree with him, but continued to hold his original position. On the other hand, most informed creationists of today do not accept Linnaeus' original opinion.

Degrees of Variation

The writer believes that most creationist biologists of today will hold that there is no one level of the biological hierarchy that can be equated with the 'kind' of creation. One should realize that not all species are alike in scope and variability. For one thing some species are extraordinarily plastic, e.g. dogs. This is evident when we consider the great variation found among them. As one considers the range in size from the chihuahua to the St. Bernard, in snout length from the English bulldog to the collie, in general build and elevation from the dachshund to the greyhound, etc., incredulity increases to the point where it becomes difficult to believe that these can all be members of the same species. Yet it is a matter of recorded history that all the breeds of dogs we have today, no matter how varied, stem from one original mongrel form.

On the other hand, in other animal forms, we find that some change or vary very little, e.g. cats. If you visualize the varieties of cats that exist today, we find that the variations seem limited to fur color and length. Examples would be Siamese, Abyssinian, Persian, wildcat, tiger, tabby, tortoise shell, bobcat, Manx, and lynx. The last two mentioned also vary as to tail length. Otherwise there is very little variation in such features as the position and shape of the ears, the length of face, and the general body form. Variations would seem to be essentially a matter of fur pattern and length. In fact, when we examine the large cats, such as panthers, jaguars, and pumas, we find that they bear an uncanny resemblance to the house cat in everything but size. These large cats will have similar ears, the same short snouts, retractile claws, and long tails.

In the plant world, we find similar examples of two such extremes in degree of variation. One can consider the ginkgos, which demonstrate very little variation in leaf form from the fossils to present leaves taken from living trees. On the other hand, the extreme degree of variation found among the leaves of oaks and hawthorns makes classification of the latter as to species almost hopeless.

Most creationist biologists also believe that any forms which can cross breed would be considered members of the same created *kind*. Frank Marsh developed this theme when he proposed that we rename the created *kind*, the *baramin* (bara, Heb. create and min, Heb. kind) instead, so that the *kind* concept will not become confused with the biologists term, *species*.[33] Today, most informed creationists do not consider the *kind* as being synonymous with any specific hierarchial level. While in some instances it could be equivalent to that of species, in others it might be equivalent to the genus rank, and in still others it might be equivalent to the family rank.

---

[33]Marsh, Frank L., *Variation and Fixity in Nature*, Pacific Press, Mt. View, 1976, p.36.

VIII. MACROEVOLUTION

The Continuum of Life

Essentially, Darwinian evolution involved small random variations. The favorable variations, i.e., those with adaptive value, were the raw material for natural selection to work on. Favorable variations aided their possessors in the struggle for existence, and thus both possessors and variations survived. Unfavorable ones were eliminated by the death of the organism. Thus a steady, gradual change resulted, i.e. the amphibians gave rise to reptiles, the reptiles to mammals, etc. The operative word is *gradualism*.

It followed, in Darwin's opinion, that there would be an unbroken progression of forms, ranging from simple to complex. The record would show a *continuum* of life, as Dobzhansky once called it. Obviously this continuum would have to include fossil forms as well as those still living. The two, living and fossil forms, meshing with each other would present the total of the continuum.

However, the evidence for this slow progression from simple to complex, which Darwin postulated, was not substantiated by the fossil evidence in his day. He recognized that here was a falsification of his theory; so he took refuge in the hypothesis of 'the imperfection of the fossil record'. In fact Darwin used this term as a heading for chapter X in his book, *The Origin of Species*. However he had faith that this lack of evidence would be remedied as time went on and more and more fossils were discovered. He hoped that ultimately there would be an unbroken succession of forms, ranging from the primary simple ones to the present day diversity.

The Loyal Opposition

Through the years that followed the publication of the *Origin of Species,* Darwin received little or no support from some of his fellow evolutionists. Such men agreed with him as to the 'fact' of evolution having occurred, but they utterly disagreed with him as to the methods by which it came about. The following are probably some of the most noteworthy.

The first, Austin H. Clark, was the curator of paleontology at the Smithsonian Institution, Washington, D.C. From that vantage point he examined the fossil evidence in his care. Seventy years after Darwin, he came to the conclusion that the fossil evidence gave little if any more support to the theory of evolution than it did in Darwin's day.[34] In his opinion the picture of evolutionary development presented a forest of

---

[34]Clark, Austin H., *The New Evolution,* Williams & Wilkins, Baltimore, 1930.

trees rather than a single tree of life. Technical references began to recognize this by the use of the terms *monophyletic* and *polyphyletic* evolution which began increasingly to appear in the technical literature. The gaps due to the lack of transitional fossils were as real as ever and had not been filled. The new terminology recognized the fact that the fossil record better demonstrated a sudden origin of the major groups of plants and animals, rather than the gradualism proposed by the Darwinians. In point of fact, Clark once wrote that 'On the basis of the paleontological record, the creationist has all the better of the argument'.[35]

Although Clark's statement was made in 1928, nothing has changed since that day. This whole situation, as well as his proposed solution is outlined in his work *ZOOGENESIS*. In essence, he anticipated Goldschmidt, when he proposed that the method of macroevolution was a sudden change in the organism as a whole from one form to another form. Clark thought that this took place at the gastrula stage in embryological development.

J. C. Willis, the director of the Botanic Gardens in Rio de Janeiro, Brazil, devoted much of his time to the study of geographical distribution. As a result of this study, in 1922 he began to recognize the weaknesses in Darwin's arguments and proposed that Darwin had actually gotten things upside down. He arrived at the conclusion that therefore Darwin's natural selection could not be responsible for evolution.

In 1940 his final thoughts were published in a work entitled *The Course of Evolution*.[36] In this work he developed his concept that macroevolution was not operating on varieties to form new species, which would then culminate in new phyla. Rather he felt that evolution had operated on the phyla, giving rise to new classes and so on *down* the hierarchy, with the species as the end result. So at an early date he felt that Darwin's natural selection was bankrupt in terms of being a cause for major changes. Actually, Fleeming Jenkin had earlier pointed out to Darwin that his theory did not account for the accumulation of new characteristics that his theory required. Instead, any that appeared would be soon lost by 'swamping out', as was known by all animal breeders at that time. Truly natural selection can only eliminate the unfit, but can never produce the fit!

In the same year that Willis' work was published, there appeared the writings of Richard Goldschmidt. He was a geneticist, who after 40 years of work with fruit flies, became discouraged with his utter lack of results in finding a genuinely new form. His experiments all resulted in

---

[35]Clark, Austin H., "Animal Evolution", *QUARTERLY REVIEW OF BIOLOGY*, December 1928.
[36]Willis, J. C., *The Course of Evolution*, Cambridge, London, 1940.

just fruit flies. Therefore he decided that the explanation of evolution as the result of a series of cumulative micromutations was false. He issued a famous challenge to all micromutationists, calling on them to postulate the series of changes which would produce the feathers of birds, the venom glands of snakes, the mammary glands of mammals, and some dozen others. His challenge was never taken up.

Goldschmidt's solution to the problem was the appearance of one great macromutation that would result in an utterly different organism—this mutation occurring in the embryo stage. For example, he proposed that a reptile once laid an egg and out hatched a bird! He readily granted that there were some difficulties in this proposition. For one, such a major change might result in the production of a monster, but he hoped that out of many such episodes would emerge one functioning form. So there would be the production of a 'hopeful monster'. However he did not solve the difficulty of arranging the production of at least two such monsters at the same time, so as to provide reproduction and perpetuation of the new form.[37]

A fourth member of this group was the noted Swedish botanist, Heribert Nilsson. Towards the end of a long career in genetics and evolution, he produced two volumes written as an evaluation of evolution by natural selection. Nilsson considered that the macroevolution of his day was refuted among other factors by the evidence for geological catastrophes and genetic stability. His evidence for catastrophism ranged from the coal beds of Germany to the amber found in the Baltic.

Let us consider what Nilsson has to say about the amber:

> "The largest deposits are found in East Prussia, where layers of 3 meters may be found. Goeppert has estimated the amber in this region (totaled out) at 5 billion kilos. It is quite obvious that these large quantities cannot have been formed on the spot. The theory is that they have their origin in resin from the conifers of the countries around the Baltic. But the conifers growing there do not normally exude resin in any quantities. It has, therefore, been necessary to postulate that heavy attacks of diseases and immense forest fires have induced the effluence of large quantities of resin. Finally, some catastrophe has thrown the forests with their lumps of resin into the sea. The amber and pieces of branches, but no stems, have then been transported by the sea to the coast of East Prussia. This is hardly an autochthonous process.
>
> In the pieces of amber, which may reach a size of 5 kilos or more, especially insects and parts of flowers are preserved, even the

---

[37]Goldschmidt, Richard, *The Material Basis of Evolution*, Yale, New Haven, 1940.

most fragile structures. The insects are of modern types and their geographical distribution can be ascertained. It is then quite astounding to find that they belong to *all regions of the earth*, not only to the Paleaearctic region, as was to be expected. Typically tropical species occur, from the Old World as well as from the New. The same is the case with the plant fragments.

Leaves of tropical trees from East India, Borneo, Australia, and South America are mixed with those from such homely shrubs as *Evonymus* and *Daphne* and weeds as *Polygonum Convolvulus* and *Geranium molle.* The genus *Pinus* is represented, but so far as it is possible to judge from the needles, none of the species are at home around the Baltic; they are Japanese or North American. *Pinus succinifera,* which should be the source of the amber, is a completely hypothetical species.''[38]

And so it goes on, plants from both tropical and cold-temperate regions found encrusted in the same pieces of amber. Certainly Nilsson felt that this was flying in the face of all Lyellian geology. Therefore he felt that he could not avoid the postulation of a catastrophe.

He also pointed out all of the mutant fruitflies produced in the gentics laboratory are always constitutionally weaker than their parent form or species. Therefore they were never found in Nature[39]. When he examined all of the available evidence, he came to the conclusion that as an explanation of macroevolution, the synthetic theory should be discarded. He proposed to substitute his 'emication' which in Nilsson's thinking consisted of new forms originating in the gamete by biosynthesis.[40] This was similar to Goldschmidt's answer, the one-step macromutation. Unfortunately, his great work, which was written in German, was never even translated for American publication.

To repeat, that biologist who is also a creationist, accepts a *limited* variation within the *created kind.* As has been pointed out, this could be equivalent to what is known as *microevolution,* e.g. Kerkut's Special Theory of Evolution. However, the creationist will not agree to the great extrapolation of microevolution, namely, that a demonstration of the general truth of *microevolution,* supports the idea that, *eo ipso, macroevolution,* follows.

Macroevolution Defined

Macroevolution is defined as evolution of the higher categories of the biological hierarchy. This is also what is commonly known as 'ameba to

---

[38]Nilsson, Heribert, *Synthetische Artbildung*, Gleerups, Lund, 1953, p. 1194-5.
[39]Ibid., p. 1186.
[40]Ibid., p. 1240-1.

man' evolution. At this point it becomes obvious that the terminology in the whole area of evolution provides a good deal of confusion. This arises because the term *evolution,* as it is usually used in high school and undergraduate college texts, as well as in popular science journals, generally means *macroevolution.* Yet it also is used to designate simple variation. Therefore in any discussion of the subject, it really is necessary to define the term *evolution* at the start, so that it is clear which meaning of the word is meant.

In this connection, one often encounters the technique of setting up a straw man for the purpose of knocking him down to the discredit of some point of view. A good example of this is the matter of the fixity of species. It is often claimed that creationists hold to the concept of the fixity of species. Then, when a case is documented where a particular species is not fixed, the cry goes forth, 'Lo and behold, creation has been disproved'. As has been pointed out, responsible creationists do not hold to the fixity of all species. The scientist who is also a creationist, holds that there are basic kinds that seem to remain intact, varying within the basic kind but not across *kind* lines. This basic kind unit is not synonymous with any specific hierarchial level. In some instances it might be equated with the species level, in some cases it might be equated with genus. In some cases, creationists will go as high as the order level. A knowledgeable creationist, therefore, will generally agree with the concept of microevolution.

In another example, creationists are often said to hold that there has been a fixity of the geographic and geologic settings on the earth from the beginning. In other words, nothing has changed on the earth's surface since the time of Creation. Apparently this erroneous concept was taught in the science classes of Darwin's day. Hence Darwin discarded creation, essentially because he had either been taught or else understood that creation concept held that the finches and tortoises found on the Galapagos Islands had been created there. Darwin found this inconsistent with what he found there, so he could no longer accept creation. Similarly this held true for animals and plants all over the world. Hence Darwin sought for another explanation for the origin of living things.

Obviously, any creationist who also believes in the Noachian deluge, would consider the idea of every aspect being fixed as nonsense. The deluge certainly would greatly have changed the geography of the earth. Therefore both creationists and evolutionists would agree that various geologic processes would be producing changes in the earth's surface since the beginning. These cumulative effects would have altered the configuration of the earth's surface with the passage of time. Therefore the creationist will be as interested in the question of geographic and geologic changes, as is the evolutionist.

There would also be a relationship between the Deluge and the distribution of plants and animals. If one subscribes to the concept of the Deluge, then all of the airbreathing animals would be distributed outward from wherever the Ark landed. So again, creationist and evolutionist alike would be interested in studying the factors involving geographic distribution of animals. Referring back to Nilsson's findings, the concept of a deluge is not so farfetched. Certainly the great amount of evidence he presents indicates that catastrophism is better supported than Lyellian uniformitarianism.

Fate of Early Arguments for Macroevolution

It is interesting that today, many of the arguments that traditionally have been cited in support of evolution have been discredited. One example would be the supposed evidence for Ernst Haeckel's recapitulation law (embryology). Haeckel was a German biologist who enthusiastically supported Darwin and became in Germany what Huxley became in England, a sort of press agent for Darwin. Haeckel formulated his biogenetic law, which stated that the embryos of higher animals in their embryological development pass through stages that resemble the forms of lower animals. In several works Haeckel supported his law with drawings of various vertebrates in various stages of their embryological development. Although the fraudulent nature of these drawings has been documented, they nevertheless still appear in many texts. They are intended to prove that 'ontogeny recapitulates phylogeny', or that each individual vertebrate in its development repeats the stages of its macroevolution. As to this law, *Ontogeny Recapitulates Phylogeny,* Ehrlich and Holm condemn it in these words:

> "this crude interpretation of embryological sequences will not stand close examination. Its shortcomings have been almost universally pointed out by modern authors, but the idea still has a prominent place in biological mythology."[41]

Another example is the existence of supposed vestigial organs, which is often cited as proof of evolution. If one were to examine the fate of the list of over two hundred vestigial organs that once were held to be present in the human body, one would find that this list has almost disappeared. Such organs as the thymus, the pituitary, the tonsils, the appendix, etc., were all once on this list. The concept of a vestigial organ implies that it is an evolutionary remnant without any present function. Since almost all of these supposed vestigial organs have been found, one after another, to have a function, the list has practically

---

[41]Ehrlich, Paul, and R. Holm, *The Process of Evolution* McGraw Hill, NYC, 1963. p. 66.

evaporated and with it that particular argument.

In this connection, an old bromide has surfaced again. It was reported recently that in Boston a child was born with a vestigial tail, with the implication that thus macroevolution was proved again.[42] Of course, it did nothing of the kind. A vestigial tail, to qualify as such, would have to include the extra caudal vertebrae. After all, this is a vital part of a vertebrate tail. Also, the whole structure would have to be an extension of the coccyx. This growth on the infant was devoid of any vertebrate, or even cartilage. It was a slender fatty tumor and was some distance away from the coccyx. The structure was easily excised. As such it has no evolutionary significance except to those desperate for any support to bolster up their theory.

A macroevolutionist might maintain that when we find animals (or plants) in nature grading from simple to complex, e.g., from protozoa to mammal, this is then proof that evolution has taken place from one-cell animal, through multicellular to complex. However, the creationist might make the point that the Creator created a multiplicity of forms, each to occupy an ecological niche as part of a great design. Both of the above trains of reasoning are logical. Which is correct? Since the evidence involved is in the nature of subjective evidence, (plants and animals don't carry labels) an argument could continue on these lines indefinitely. Austin H. Clark once wrote:

"It is almost invariably assumed that animals with bodies composed of a single cell represent the primitive animals from which all others are derived. They are commonly supposed to have preceded all other animal types in their appearance. There is not the slightest basis for this assumption beyond the circumstance that in arithmetic—which is not zoology—the number one precedes the other numbers."[43]

It is often claimed that classification proves macroevolution. Actually the reverse is true. If macroevolution were true, then we should find that organisms grade continuously from one form to another. The fact that it is possible to group plants and animals into discrete groups, that the whole assemblage of living forms is essentially an array of arrays is a very powerful argument against macroevolution.

Space precludes a more complete discussion of all the arguments that have been advanced to support macroevolution. However, a number of creationist works listed in the bibliogrpahy treat this whole question in more detail.

---

[42]Oakland Press, *Boston Baby Born with a Two Inch Tail*, Thursday, May 20, 1982.
[43]Clark, 1930, pp. 235-6.

IX. READING THE FOSSILS

The Paleontological Evidence

In the past, much time has been wasted in the production of unverifiable family trees. At one time, many works on macroevolution, as well as most on paleontology, were filled with family trees or phylogenies. It is interesting to note that at the present time, it is agreed that such phylogenies are by little sound evidence. As a result, the use of phylogenies has fallen into such disrepute, that we rarely find a text that still contains them. The construction of a phylogeny implies the existence of hard evidence, e.g. the finding of transitional forms as fossils. Among professional groups, when challenged, macroevolutionists will remain silent about the existence of the transitional forms which supposedly furnished the factual basis for such trees. Actually, several noted authorities recently have frankly confessed that they really don't know of any transitional forms.[44] What is extremely odd, however, is that in public appearances, as well as in writing for popular journals, one still finds macroevolutionists maintaining that hundreds of transitional forms are well documented from the fossil evidence.

Certainly it would seem, according to Corner, that the paleontological evidence supports creation more than evolution.[45] Not only did Darwin himself admit the deficiencies in the fossil evidence concerning macroevolution, but Darwin's 'bulldog', T. H. Huxley had this evaluation to give:

> "In answer to the question 'What does an impartial survey of the positively ascertained truths of paleontology testify in relation to the common doctrines of progressive modification' I reply: it negatives these doctrines for it either shows us no evidence of such modification or demonstrates such modification as has occurred to have been very slight".[46]

The Situation Today

Certainly the October 1980 meeting at the Field Museum of Natural History[47] was almost a replay of Huxley's statement. However, it would seem as if the macroevolutionists have such a profound dislike of the concept of a Creator, that they will continue to accept the theory of evolution, no matter how far-fetched, no matter what evidence is found

---

[44]Lewin, 1980, pp. 883 ff.
[45]E. J. Corner, 1961, p. 97.
[46]Huxley, Thomas H., *Lay Sermons, Addresses, and Reviews,* Appleton, NYC, 1879, p. 225.
[47]Lewin, 1980, p. 883.

"negativing" macroevolution. This attitude would seem to support Norman Macbeth's point that evolution has become a religion in our day.[48]

In addition, no matter what the evidence that is reported, macroevolutionists always have been able to adjust their theory to accommodate it. Gertrude Himmelfarb drew attention to this state of affairs when she wrote of the Piltdown Hoax as follows:

> "Nor can it be maintained, as some Darwinists have done, that the exposure of Piltdown man leaves them no better and no worse off than they were before. It does, in fact, weaken their position in regard to both their theory and their methods. The zeal with which eminent scientists defended it, the facility with which even those who did not welcome it managed to accommodate to it, and the way in which the most respected scientific techniques were soberly and painstakingly applied to it, with the apparent reuslt of confirming both the genuineness of the fossils and the truth of evolution, are at the very least suspicious."[49]

The Return of Paley

Dr. William Paley published a book in 1802 entitled *Natural Theology*. It seemed to Paley that wherever he looked, he saw the hand of a Creator. To him, design was apparent everywhere. Darwin was well acquainted with Paley, in fact he took this book along with him on the voyage of the H.M.S. *Beagle*. But although Darwin relied on natural selection as the main force in the origin of species, Darwin could not shake off Paley. In spite of all, Paley still had a valid point when he said that the whole of creation cries out "order and design". Paley's position was that just as the telescope implied the existence of the optics maker, so the existence of the human eye, implied the existence of the Creator. Yet Darwin, in spite of all his beliefs and writings to the contrary, frankly confessed that sometimes he woke in a cold sweat thinking of the human eye. Darwin was frequently conscious of the effectiveness of Paley's arguments. In fact he wrote to Asa Gray in 1860,

> "I remember well the time when the thought of the eye made me cold all over, but I have got over this stage of the complaint, and now small trifling particulars of structure often make me very uncomfortable. The sight of a feather in a peacock's tail, whenever I gaze at it, makes me sick."[50]

---

[48]Macbeth, 1971, pp. 124 ff.
[49]Himmelfarb, Gertrude, *Darwin and the Darwinian Revolution*, Chatto and Windus, London, 1959, pp. 310-311.
[50]*Life and Letters of Charles Darwin,* (3 Vol.), Murray, London, 1887, Vol. 2, p. 296.

It would seem that Darwin despite all, still felt that Design was the better argument for the existence of the eye. In his section on "Organs of Extreme Perfection and Complication' in the *Origin of Species,* he began with these words:

> "To suppose that the eye with all its inimitable contrivances for adjusting the focus to different distances for admitting different amounts of light, and for the correction of spherical and chromatic aberration, could have been formed by natural selection seems, I freely confess, absurd in the highest degree."[51]

More recently, in commenting on this statement of Darwin's, Stephen Gould remarked:

> "Natural selection has a constructive role in Darwin's system: it builds adaptation gradually, through a sequence of intermediate stages, by bringing together in sequential fashion elements that seem to have meaning only as parts of a final product. But how can a series of reasonable intermediate forms be constructed? Of what value could the first tiny step toward an eye be to its possessor? The dung mimicking insect is well protected, but can there be any edge in looking only 5% like a turd?"[52]

---

[51] Darwin, 1971, p. 167.
[52] Stephen Jay Gould, *Ever Since Darwin,* Norton, NYC, 1977, p. 104.

X. FOSSIL MAN

Human Origins

When it comes to the matter of human origins, the writer long has believed that creationists have to shed their old notion of Adam and Eve as representing the best looking among WASPs, (White, Anglo-Saxon, Protestant) with the implication that were they alive today, they could win a Mr. and Mrs. America contest hands down. An examination of all the evidence concerning fossil man indicates that Adam and Eve could more accurately be pictured as being a genetic mixture of all races and forms of man that ever have existed, both past and present. So to be consistent, in attempting to picture the possible appearance of our first parents, we also should include the features of all fossil *Homo* forms in any such illustrations.

Human Variation

Man has certainly never varied to the extent that we find among dogs, but he has varied. I don't believe that the implications of this factor have been realized to the extent that they should be. For one thing, modern man is quite variable as to cranial volume. For example, two great men of letters, Dean Jonathan Swift of England and Anatole France of France, varied by almost 1,000 cubic centimeters in their cranial volume. Incidentally, it should be noted that this is a greater difference than that between the average modern man (1400 cc.) and early *Homo* forms (900 cc.). Man has varied with respect to brow ridges from almost non-existent to those that are quite pronounced. He has varied from a pronounced chin to none. In the past he has had an occipital bun at the back of his head. At one time he possessed an inca bone in the top of his skull. He apparently has varied in the size of nasal openings. On the other hand, man seems to be consistent as regards the position of the foramen magnum as well as in having a parabolic tooth arch. He shows a consistently greater cranium to face ratio. He can and does vary in the degree of overall mandible buttressing.

Homo Characteristics

It has often been said that man is an animal. Rather it should be said that man is a thinking animal, capable of abstract thought. Man is also capable of laughter and tears. He has the capacity to appreciate the beauty found in the world about us. We see it in plants, in the animal world, and in the scenery that surrounds us so often, e.g. the mountains, the sky, etc. In his discussions on sexual selection, Darwin believed that his theory stood or fell on this matter of beauty. He said that if it could be proved that the beautiful patterns found in so much of the animal world were created to delight man or the Creator, such proof would be absolutely fatal to his theory. If the macroevolutionist exhausts himself in efforts to explain utility in the world, then beauty will rise up to confound him. It also has been said that the animal knows, but man knows that he knows. Unfortunately, the ability to reason is not capable of being fossilized.

It is a fact that among animals, man most closely resembles the apes. However, resemblances do not prove descent. Most texts will emphasize the resemblances between man and ape. But the major differences between them are almost never discussed. Some of the differences between man and ape are as follows:

Man	**Ape**
1. Large vaulted cranium	Flattened cranium
2. Mastoid process prominent (See Fig. I, A1)	Mastoid process absent or inconspicuous
3. Canines project little, if at all	Projecting canines
4. No diastema in upper jaw (See Fig. I, C1)	Disastema present (See Fig. I, D1)
5. No simian shelf	Simian shelf present
6. Lips prominent	Lips extremely thin
7. Vertebral column in 3 curves	Vertebral column in 2 curves
8. Short neural spines on cervical vertebrae	Long neural spines on cervical vertebrae
9. Relatively short arms	Relatively long arms
10. Lower-placed nipples	Higher-placed nipples
11. Female with persistent breasts	Female with flattened breasts
12. Body relatively hairless	Body relatively hairy
13. When present, body hair most prominent on ventral (belly) surface	Body hair most prominent on dorsal surface
14. No baculum (os penis)	Baculum present
15. Deep bowl-shaped pelvis	Shallow flattened pelvis
16. Bulging gluteus maximus	Flattened gluteus maximus
17. Linea aspara present on femur (roughened ridge for muscle attachment)	Linea aspara absent
18. Feet different from hands	Feet very similar to hands[53]

---

[53]Wilbert H. Rusch, *Human Fossils*, Chap. 9, *Rock Strata and the Bible Record*, P. A. Zimmerman, Ed., Concordia, St. Louis, 1970, pp. 144-5.

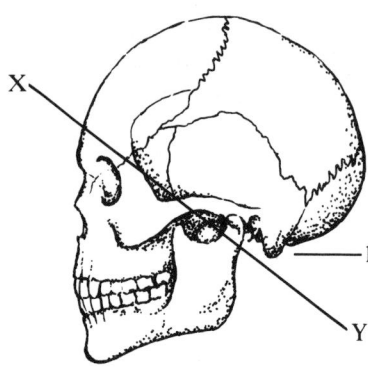
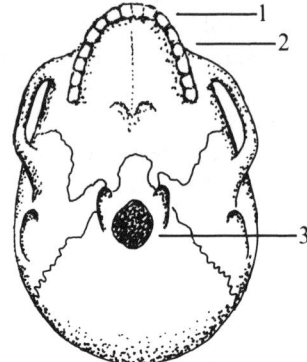

A. Human skull (lateral view)
  1. Mastoid process

C. Human skull (ventral view)
  1. Diastema missing
  2. Parabolic tooth arch
  3. Foramen magnum centered

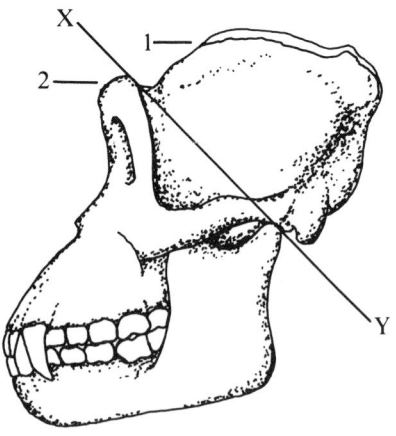
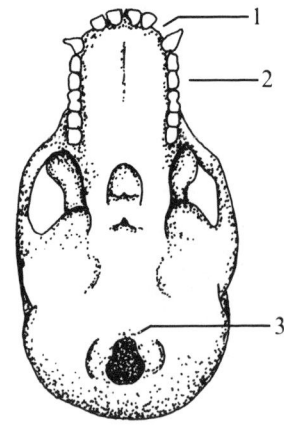

B. Gorilla skull (lateral view)
  1. Sagittal crest
  2. Supra-orbital ridge

D. Gorilla skull (ventral view)
  1. Diastema present
  2. U-shaped tooth arch
  3. Foramen magnum not centered

Fig. I — Skull of man (A and C) compared to skull of ape (B and D).

Note the greater proportion of the cranial vault (above the line X-Y) to the face (below the line) in man compared to the gorilla. (Ignore the lower jaw in this composition.)

It should be noted that items 6, 10, 11, 12, 13, and 16 are not subject to fossilization. Therefore the determination of the humanity of a given find may become difficult. This is particularly true when all that is available is the skull or skull portions.

In terms of man's mental capacities, two items are used as criteria. The first is the controlled use of fire. Since no known animal uses fire in any way, anthropologists use this item as a definite criterion of the humanity of a given fossil. All animals fear fire, but man uses it for warmth, the preparation of food, and protection.

Secondly, man is the only organism that not only makes a tool, but takes the further step of also making a tool whose only function is to enable him to make another tool. This is an abstraction that takes an organism beyond the activity of simple tool use. Chimpanzees recently have been observed to use leaves as a sponge to soak up water from a cavity. But that apparently is as far as they can go. Man posseses the capability to make items which have no use in themselves except to be used to make another tool.

The Human Fossil Evidence

The scientist who is a creationist must take all available *factual* evidence into account and deal with it. That would include the type and extent of the variations mentioned in the previous paragraph. It also would include all the various fossils of humans that have been discovered. However, it should be recognized that the making of reconstructions of the soft parts of these fossils is not factual, but very speculative, since such parts are not preserved. For that reason none of the available reconstructions should be considered as valid evidence. Rather they are little more than an expression of one man's or a group of men's opinion. The writer once ordered two reconstructions from Skhul V, one of the Mt. Carmel skulls, which were executed by two different anthropologist sculptors, operating under different anthropologists' ideas. When they arrived, the busts were so dissimilar that they didn't even look like brothers![54]

When one considers the total picture of human fossils, one finds that there is an astonishingly small amount of material available. Relatively few complete skeletons of early humans have been found. Some of the early supposed ape ancestors of man, e.g. *Sivapithecus indicus* are represented only by a mandible fragment.[55] The writer was once in attendance at an Academy of Science meeting, where a noted anthropologist was heard to make the remark that 'we have plenty of competent anthropologists, what is needed are more competent fossils'.

---

[54]Rusch, *Ibid.*, pp. 156-159.
[55]Rusch, *Ibid.*, p. 154.

It would be possible to place *all* hominid fossils, prior to Neanderthal forms, on the tables of the usual laboratory. Most fossil finds are of skulls and skull parts. In far too many cases, these are limited to jaw fragments. Sometimes these are the only concrete evidence for the existence of a given form. Yet even with such a general paucity of fossils, in many instances one can arrive at a somewhat coherent concept of a particular form.

The writer agrees with those who consider the Australopithecines as being definitely ape-like forms.[56] One of the characteristics of this form is the V-shaped angle of the lower mandible and the heavy buttress or superior mandibular torus at the point of the jaw. This buttress is not the same as the simian shelf, which is much smaller and is known as the inferior mandibular torus. In the apes, we find a U-shaped dental arch and a very pronounced buttress, in addition to the considerably smaller projection called the simian shelf. In humans we find a distinctive parabolic outline of the dental arch as well as a lack of either buttress or simian shelf.

---

[56]McHenry, H. M., "*A View of Hominid Lineage,*" *Science,* Vol. 189 p. 988.

XI. THE LATEST FINDINGS

Lake Turkana Finds

A very interesting picture of human history is developing according to the finds of Richard Leakey and others. *Homo* or man seems to occur consistently in the same strata as his supposed ancestors, the Australopithecines. Obviously if humans had evolved from Australopithecine forms, then the descendant should not have co-existed with his ancestor. The latest examples of this sort of evidence are the two Leakey finds, identified as KNM-ER 3733 and KNM-ER 406.

These have been discovered in the Lake Turkana (formerly known as Lake Rudolph) area in East Africa.[57] One of these finds, known as KNM-ER 406, is almost a twin to that Australopithecine known as Swartkrans. Both are definite *Australopithecus robustus* forms. On the other hand, KNM-ER 3733 is clearly a *Homo* form. Yet both of these were found in *situ* in the same stratum, called the Upper Member of the Koobi Fora tuff. This evidence indicates that there is little room for doubt of the contemporaneity of *Homo erectus* and *Australopithecus* at an early date, e.g. Pliocene.

The same situation also occurs in Swartkrans, South Africa. Here a *Homo erectus* form (originally named *Telanthropus capenensis*) was found in the same stratum as an *Australopithecus robustus* type.[58] Finds like these, where both ancestral and descendant forms are found under circumstances indicating that both lived as contemporaries, make the various evolutionary scenarios of man very difficult to maintain. In addition, there is the fact that there is little agreement among anthropologists as to lineages or sequences of the various members that make up the ancestral trees.

To sum up, in 1970, when the book *Rock Strata and the Bible Record,* was published, the writer wrote as a closing paragraph to his chapter on Human Fossils. It read as follows:

> "There is preliminary evidence that a species of *Homo* coexisted with *Australopithecus.* If this is substantiated, it will show that the genus *Homo* is as old as the oldest known hominid remains. The differentiation of species within *Homo* is dubious. There are living men with most or all of the skeletal characteristics of *Homo erectus.* The fossil evidence supports the conclusion that species of *Homo* similar or identical to *Homo sapiens* have

---

[57] R. E. F. Leakey, "*Australopithecus, Homo erectus,* and the Single Species Hypothesis," *Nature,* Vol. 261, June 1976, pp. 572-5.

[58] Paul L. De Vore, Ed., *The Origin of Man,* Werner Gren Foundation, 1965, p. 11.

existed since the initial appearance of humanoid fossils in the geological record. Therefore it may be concluded that fossil evidence offers no support for any scheme of the evolutionary descent of man, either within hominid genera or from primate ancestors."[59]

The discoveries in the Koobi Fora formation have occurred since that paragraph was written, and have substantiated the statement made at that time.

'Lucy'

In addition to the finds of Richard Leakey, Donald C. Johanson has discovered a form commonly known as 'Lucy'. This find was made in the Afar triangle of northeastern Ethiopia, in strata that were dated at 3.6 million years. When first discovered, Johanson hailed his find as being the common ancestor of both *Homo* and *Australopithecus*. Johanson determined that Lucy was to be classified as *Australopithecus afarensis,* man's oldest ancestor. A recent summary of the discussion concerning 'Lucy' is found in *Science News*.[60] The importance of this particular find has dwindled considerably due to the fact that the age of this particular fossil has been revised to a range of 2.9 to 3.2 million years B.P. This revision puts 'Lucy' into the same range as previously discovered *africanus* forms found in South Africa.[61] Therefore we find that 'Lucy' has now been reclassified as being in the same group as other *A. africanus* forms which have been found in other localities, but still strata of the same age. It seems as if among some anthropologists, classification depends on the age of the stratum in which the fossil is found, instead of on morphology! Obviously this procedure could effectively dispose of any fossil evidence that might be used to question the validity of macroevolution. Certainly this procedure flies in the face of all taxonomical principles that the writer has encountered in his biological career. It brings to mind a criticism that George Gaylord Simpson once made of the nomenclature situation he encountered in the field of anthropology. His evaluation was as follows:

> "Insofar as the chaos is due to faulty linguisitics rather than to zoological disagreements, it stems either from ignorance or from refusal to follow rules and usages. This must be the only field in science in which those who do not know and follow the

---

[59]Rusch, *Ibid.,* pp. 171-2.
[60]Herbert Wray, "Lucy's Family Problems," *Science News*, Vol. 124, July 2, 1983, pp. 8-11.
[61]Michael H. Day, "Lucy Jilted?", *Nature* Vol. 300, 16 December, 1982, p. 574.

established norms have so frequently had the temerity and opportunity to publish research that is, in this respect, incompetent."[62]

Granting the validity of the basic groups, *Australopithecus africanus, Australopithecus robustus, and Homo erectus,* then an alternative scenario would consider that all finds could be considered as examples of variation within the species or microevolution. When that scenario is considered, then much of the difficulty presented by the 'human' fossil material ceases to cause such great difficulty for the creationist. Particularly is this true when one considers the consistent evidence of various *Homo* finds turning up in the same stratum as the *Australopithecus* finds.

---

[62]George Gaylord Simpson,"The Meaning of Taxonomic Statements", *Classification and Human Evolution,* Sherwood Washburn, Ed., Aldine, Chicago, 1963, p. 24.

XII. THE AGE OF THE EARTH

The Age of the Earth

Many individuals consider the age of the earth a question which is integral with the subject of origins. The writer of this article disagrees with this point of view. Creation by its very meaning deals with the question of origins: chance or Creator? This can be handled without *necessarily* determining the date of such origins. This is borne out by the fact that noted and articulate creationists have varied greatly in their estimate of the date of creation.

Such estimates have ranged all the way from Ussher's opinion, namely 4004 B.C. At the other extreme we have Douglas Dewar and L. Merson Davies who proposed the base of the Cambrian Period as being the time of creation. This is generally held to be about 500 my B.P. (million years before present). We have individuals who hold that 10,000 years is the maxium allowable age of the earth. One can agree on the origin of life by creation without coming to an agreement as to when such a creation took place.

Unfortunately, among the latter group there are those who insist on the acceptance of that limitation by anyone who would wish to be considered a creationist. This is not only unfortunate but also presumptuous. This is strictly their own belief, and it is not their right to determine who is and who is not a creationist. A case in point: Douglas Dewar, who was previously referred to, was a noted British creationist, who was most effective in debates (with J. B. S. Haldane for one) as well as in his many writings against macroevolution. He served in this manner in England during the 1930's and 1940's. Yet he held to the whole geological timetable and believed that the Cambrian period was the time of creation.[63] But to deny him his place among effective creationists is really rather presumptuous. Dr. L. Merson Davies, Gold Medalist in Geology, shared Dewar's views. Again, one might mention another noted creationist author, Byron Nelson, the author of *After Its Kind* and other similar works. In his last work, *Before Abraham,* he developed the thesis that Adam dated back to 900,000 B.P. Other creationist writers could also be quoted. So it becomes apparent that among creationists there are considerable differences of opinion on this matter of age.

The Weaknesses of Radiogenic Dating Methods

Nevertheless, the writer would like to point out that none of the

---

[63]Dewar, Douglas, *The Transformist Illusion,* DeHoff, Murfreesboro, Ind., 1957.

radiogenic dating methods in use today are accurate or precise. As has been said, creationists vary considerably in their opinions as to the time of creation, and they can live with short or long periods of time. However, a macroevolutionist *absolutely requires* long periods of time. There has been little change in this point of view since Darwin's day when he bemoaned the fact that Lord Kelvin only gave him 20 million years or so for the age of the earth. Darwin agreed that this estimate posed a very serious problem.[64] At another point in the *Origin* he stated that anyone who didn't accept Lyell's long periods of time, might just as well stop reading his book right there.[65] It is remarkable that the age of the earth seems to be increasing about 1 billion years every generation. Apparently it is a figure 'subject to change without notice'.

One frequently encounters a statement that seems to have been promoted to the status of a law among macroevolutionists, namely *given enough time, the impossible not only becomes probable, but inevitable!* What a totally vacuous statement! It sounds profound, but it obviously is utterly meaningless. It certainly is not science, since it is not testable. (Popper dealt with this when he wrote "A theory which is not refutable by any conceivable event is non-scientific"[66]) Yet many people are uncritical enough in their thinking to be taken in by this statement so that they consider it a valid argument. Actually, the statement suggests that time has become the great *deus ex machina* of macroevoluton. We have 'time' proclaimed as the great creator that has replaced the Creator. *Given enough time, anything can happen!*

This question of the age of the earth is a very complex one. The whole matter rests upon one's assumptions. But by definition, assumptions are statements accepted without proof. The writer does not accept the radiogenic systems of dating as valid for several reasons. One of them is the aforementioned increase in age per generation. Ignoring this point ignores the question that if the data, which gave rise to the previous dates were faulty, what becomes of this data? And how valid are those based on any succeeding data? And what figure will be given as the age of the earth 100 years from now?

In the second place, it is not always realized that every radiogenic date rests on its own series of assumptions. These are as follows:
    a) The radioactive decay rate is a constant, not affected by any physical or chemical action.
    b) The rock or mineral has been a closed system since its formation; that is there has been no loss or gain of the parent

---
[64]Darwin, 1971, p. 442.
[65]Ibid., p. 294.
[66]Karl R. Popper, *Conjectures and Refutations*, Routledge & Kegan, London, 1963.

radiogenic materials other than due to radioactive decay.
c) After the time the radiogenic clock was set, there should have been no occurrence capable of resetting the clock.
d) The time of formation of the rock or mineral is short compared to its age.

If any one of the assumptions should be invalid, it follows that the date determination based on them also is invalid, regardless of the precision of any measurements taken. No matter how precise the measurements are, they have little to do with the accuracy of the result in any age determination. It is the question about the validity of the assumptions that is no doubt responsible for any discrepancies. It is factual that the records of dating estimates show that there are numerous anomalies encountered in the results achieved with all methods used in radiogenic dating estimates.

One example is found in some lava flows off the coast of the island of Hawaii. The Kaupulehu flow of Hualalei was known from historical records to have cascaded into the sea about 1801 during an eruption. However when a researcher subjected the eight samplings retrieved from the solidified lava on the ocean bottom at that locality for a dating check, the K/Ar (potassium argon) method gave postulated ages for the lava as follows:[67]

1) 160 million years
2) 791 million years
3) 960 million years
4) 1,500 million years
5) 791 million years
6) 2,040 million years
7) 2,470 million years
8) 2,960 million years

Two points can be made here. In the first place there is the large difference between the time of the flow, and the postulated time given by radiogenic methods. Then there are also the large discrepancies between dates of samples from the same flow. The explanation advanced by one author is that a) the lava was extraordinarily rich in zenolithic inclusions to begin with, and b) these xenoliths contain large amounts of Ar-40.[68]

But these explanations do not seem adequate to account for both points above. As per usual, all kinds of adjustments are made to bring

---

[67] John G. Funkhauser and John J. Naughton, "Radiogenic Helium and Argon in Ultramafic Inclusions from Hawaii", *Journal of Geophysical Research*, Vol. 73, 14, 1968 pp 4601-07.

[68] Dalrymple, G. A., and J. G. Moore, "Agron 40: Excess in Submarine Pillow Basalts from Kilauea Volcano, Hawaii", *Science*, Vol. 161, pp. 1132-1135.

the results in line, including the postulation of the escape of argon gas. It is factual that an eruption cascading into water will produce the phenomenon known as pillow lavas. However, in my reading, the K/Ar dates assigned pillow lavas are always considered to be correct with never a suggestion that any such dates might be as spurious as these Hawaiian ones, and for the same reason. Many other examples of anomalous radiogenic dates could be mentioned. It is now admitted that 50% of uranium dates of one sort or another are anomalous. One could also discuss the fact that there is more helium in the world than the standard radioactivity theory allows.

Finally, there is a built-in extrapolation of such magnitude that only wishful thinking would seem to excuse it. It lies in this, that we have observed radioactive decay for only some 100 years, since before then the phenomenon was generally unknown. Yet on the basis of such short experience with direct observation, we find extrapolations for more than a million, yes even a billion years. There would seem to be an inherent fallacy here. As proof for the correctness of the reasoning it is often claimed that the U.S. Coast Guard indulges in the same type of extrapolation when it produces its tide tables. But although the U.S. Coast Guard has observed tide phenomena for some 200 years, they do not claim therefore that they can produce accurate tide tables for 15,000 years in the future. Neither do they claim that they can produce tables for some 6,000 years in the past, which would give the tide for, say 4104 B.C. Obviously the Coast Guard would never be that foolish. Extrapolation is only valid with a long base of data followed by only short extrapolations in each direction.

The Evidence for a Young Earth

Not to be overlooked is the fact that there is also evidence that can be used to support a relatively short age for the earth. Thomas Barnes has reported on the fact that the decay rate of the moment of force of the earth's magnetic field has been determined from the time of Gauss. Comparing the change in value from then to that at present provides the data for an extrapolation to the past and the future. Such data would seem to indicate a 20,000-year value for the age of the earth[69]. Admittedly this is also acting on an assumption that the decay rate remains constant. But the extrapolation is not of the fantastic range as that used in radiogenic estimates.

Even stronger evidence for a relatively young earth is indicated by the

---

[69]Barnes, Thomas G., *Origin and Destiny of the Earth's Magnetic Field*, Tech. Mon. #4, ICR, San Diego, 1973.

studies made by Robert Gentry on the pleochroic halos. These are radioactive halos, so named since they are circular patterns found in some minerals. These halos are due to the radioactive emissions of uranium, thorium or polonium. Because of the radioactive emissions, a discoloration is produced that extends equidistant in all directions. Then when the rock containing this material is sectioned, the discoloring shows as a halo around the radioactive nucleus. The main problem lies with the halos of polonium. This problem exists because the radioactive isotopes have half-lives measured in days. The question arises when these are found in pre-Cambrian granites. No one is quite sure how granites are formed. But it is supposed that this is a slow process, taking immense periods of time. Obviously there is a great discrepancy here. The existence of polonium halos in granite says that the granite was formed within a short time, yet the dating methods all say that it must have been millions of years. Here we have a case of experimental evidence indicating one theory, and the applications of numerous assumptions indicating a diametrically opposed theory. These facts have been reported adequately in the AAAS journal, *Science*.[70] Gentry's work has also supported catastrophic episodes in the past by the demonstration of the presence of uranium halos in coalified wood, as well as his studies on lead and helium found in zircons present in pre-Cambrian granite. The former has implications with regard to the existence of a massive catastrophe.[71,72]

It is often said that creationism has no predictive value. The writer feels that this is nonsense. A macroevolutionist might predict, on the basis of his theory, that one should be able to determine the point at which $t = 0$, that is how many years ago the earth originated. He should be able to find that all data point to that one value, and then he has solved the question. The writer, as a creationist, would postulate that the result of a creation was a 'breaking into the time cycle', with everything suddenly a going concern. This would imply that the earth looked as if it had been around a long time, but actually, shortly after creation, everything was relatively young. So the writer would predict that all attempts to determine $t = 0$ would end in failure, since there would not have been an evolutionary development of anything. Now it would

---

[70]Gentry, Robert V., et al, "Radiohalos in a Radio-Chronological and Cosmological Perspective," *Science*, 5 April, 1974, Vol. 184, pp. 62-66.

[71]Robert Gentry, "Radiohalos in Coalified Wood: New Evidence Relating to the Time of Uranium Introduction and Coalification," Science," Vol. 194, 15 Oct. 1976, pp.315-8.

[72]Robert Gentry, "Radioactive Halos," *Annual Review of Nuclear Science*, XXIII, 1973.

seem that the writer is on sounder ground, since so many dating results are discrepant. For example, in one case, various radiogenic dating methods give an age of 440 my, 770 my, and 1040 my, as the age of the same rock![73] This would certainly weaken any claim towards the establishment of t = 0! Yet these results would seem to be more consistent with the writer's proposition.

At the same time the writer wishes to be fair about this matter of age of the earth. He is reacting to the situation he finds where despite numerous disclaimers as to certainty of these measurements, yet most texts and works on the subject give a definite impression that the vast ages are proven beyond doubt. Furthermore, anyone who doesn't accept the concept of the billions of years is considered as not being quite all there. The writer has tried to indicate that this absolute certainty is just not justified.

But neither does he maintain that there are no problems for the creationist in this matter of age. The creationist cannot prove his point either. There are some difficulties for him as well. For example, the layers of sedimentary rock formations found below the surface from the Tennessee line in the north to Horn Island in the Gulf of Mexico are well documented because of the search for oil and the drilling of many wells, both successful and unsuccessful. Maps are available for the cross section of this area. These maps also show salt domes that have risen to the surface, piercing overlying layers of sediments as they rose. A study of this section will pose numerous time problems for the recent creationist. Another problem area is the occurrence of numerous coral reefs, both in the Bahamas in the Atlantic, as well as the Marshall Islands in the Pacific. The vast depths of these reefs pose the problems. Finally, although there are numerous anomalies in age determinations, there still seems to be a consistency in the matter of results of great magnitudes. It is unfortunate in age discussions, any evidence for a relatively young earth is so quickly written off and ignored. This does not convey confidence in the objectivity of any discussion on the matter of age.

So as with macroevolution, the matter resolves itself into — what does one believe? What are one's basic assumptions? Which evidence does one choose to accept and which to reject? The evidence is not unequivocal either way. But that is exactly the reason why students should have freedom of choice. One can say there are problems with a belief in creation, but then there just as many if not more problems with macroevolution.

---

[73]Harold F. Blum, *Time's Arrow and Evolution,* Princeton Un. Press, Princeton, 1951, p. 10.

XIII. CONFLICTS AND CONTRADICTIONS

The Creationist Parent's Dilemma

The dilemma that the creationists, particularly parents, find themselves in today is that students of elementary through high school age are compelled by law to attend school. Perforce then, having the advantage of being supported by the tax dollar, in many regions of our country, there is no other choice for them than attending the public school. The curricula of these schools are usually controlled by state departments of education. These in turn often are greatly influenced by the state universities. The result is that the majority of students at a public school are subjected to a consistent indoctrination in macroevolution from the lower elementary levels on. A perusal of science texts indicates that children are apt to be taught insidiously and gradually that macroevolution and natural selection are demonstrable scientific facts. Further, that macroevolution is as factual as any concepts in the fields of physics and chemistry are said to be. This indoctrination reaches a climax in biology and earth science courses. This state of affairs has been amply documented by the work of the Mel Gablers in Texas.[74]

Results In the Public School

As a result, whatever their religious beliefs may be, in their science classes (and others), students are exposed to the guiding principle that generally only those concepts that can be proven are to be accepted. And of course, evolutionism is considered to be one of those. Obviously such a philosophy comes into conflict with the beliefs of people of a number of religions, e.g., conservative evangelical Christians, orthodox Jews, as well as orthodox Muslims. It follows, that any student who has been raised to accept creation as a matter of personal faith will then experience an inner conflict as he or she is subjected during the school day to a relentless and constant attack on that faith. Yet it is never even mentioned that evolution requires just as much faith. Futhermore, those students who are creationists will often find themselves ridiculed or even discriminated against for their beliefs. This, of course, becomes a flagrant violation of those students' civil and constitutional rights!

Level of the Problem

At this point the reader should be reminded that primarily this argument is directed against that level of public education where attendance

---

[74]James C. Hefley, *Are Textbooks Harming Your Children?*, Mott Media, Milford, 1979, pp. 43-51.

is *compelled* by law. At the level of colleges, universities and the like, attendance is not compulsory. Here any student who decides to continue beyond high school has a choice which institution, if any, to attend. So colleges and universities are not considered here.

But on the lower levels, the only way a student may escape attending a public school, is to attend a private school. This is possible only if the parents have the means to pay the tuition charges of the private school in addition to supporting the public school system through their taxes. The writer believes that it is precisely over this point that the present conflict arises. Harassed parents feel utterly frustrated, when they view the often humanist nature of their children's education in many public schools. They feel helpless in the face of an utter lack of options in their choice of education. To ensure not only the intellectual but also the spiritual protection of their children, they support the enactment of various Balanced Treatment Acts, which are the subject of so much controversy today. Such laws, they feel, provide the only solution to their dilemma.

A Possible Solution

It is also important at this point to realize that what is called public school education did not exist at the time the founding fathers wrote the Constitution. In those days, the whole of education was carried out by private schools, either church or secular. Incidentally, this state of affairs did not seem greatly to distress the founding fathers. Neither in the Declaration of Independence nor in the Constitution is there any mention of education. Applications of the latter to the public school system therefore are extremely questionable, since the public school concept did not come into widespread practice until the latter half of the 1800's. It is not that the founding fathers were not concerned about education, they apparently felt it to be a local and private matter, and therefore not in the province of the federal government.

The writer feels the solution is to enforce the following requirement; at the levels of *compulsory* attendance, at least the existence of an alternate theory of origins must be recognized. A further requirement would be that the utterly theoretical nature of macroevolution be taught. The real problem is how to bring this about. One of the difficulties is that macroevolution is held by many of its proponents with a sort of religious fervor, and they feel a call to make as many converts to it as possible. Macbeth has made this clear in his work.[75]

At the same time, a perusal of the quotes in Appendix B will reveal that there is a minority of scientists, who although macroevolutionists,

---

[75]Macbeth, 1971, p. 124 ff.

still tell it like it is. They are frank in dealing with the difficulties of macroevolutionary theory and say so in print. Unfortunately, a number of macroevolutionists apparently believe that quoting of these individuals is unfair and dishonest. However, the writer has always felt that favorable evidence elicited from a hostile witness is choice evidence indeed and very legitimate. But the existence of negative evidence and doubts regarding evolution is never hinted at in the lower-level textbooks. Hence the student receives a biased view of the whole matter of origins. Therein lies the greatest problem and cause of concern.

The writer is often at loss to understand why in a scientific discussion, all rules of logical thinking are supposed to be suspended. Therefore he feels that it is right and proper to draw the attention of the scientific layman to the various disagreements within evolutionism's ranks. Proponents of macroevolution claim that although there is great disagreement over the *methods* whereby evolution occurred, yet all scientists are agreed that the fact of evolution is as well established as the law of gravity. It doesn't seem to occur to them that this inability to determine how evolution worked might be due to the fact that it *never occurred!* It would seem then that in the matter of macroevolution, we have, in the words of Dr. W. R. Thompson, an instance of numerous scientists —

"rallying to the defense of a doctrine that they cannot define scientifically. Certainly they cannot demonstrate it in a scientific fashion. Nevertheless they attempt to maintain it in public by suppression of criticism, specious arguments, and by the elimination of its difficulties. This certainly is not good science."[76]

---

[76] W. R. Thompson, "Introduction," *Origin of Species,* Dutton, NYC, 1956, p. XXIII.

XIV. CONCLUSION

It is the opinion of this writer that when one considers all of the points cited in the body of this paper, macroevolution becomes an untenable position.

It should be emphasized that a biologist who is also a creationist accepts a *limited* variation within the *created kind*. As has been pointed out, this could be equivalent to what is known as microevolution, i.e. Kerkut's special theory of evolution. The creationist has little to quarrel with in much of the evidence for microevolution. But this does not mean that the theory of macroevolution is validated. Kerkut pointed out the weakness of macroevolution, not even granting it the status of a theory, but limiting it to a working hypothesis at best. Others have continued to point this matter out in recent times. The writer then repeats his earlier assertion that there are two explanations of origins, creationism and macroevolution. Neither can be proven, both must be taken on faith.

Finally, this whole matter of origins, including which theory the student is to accept, should be a matter of choice on the part of the students and teachers, rather than a one-sided brainwashing. Especially should this hold true in the public elementary and secondary schools where attendance is compulsory. Macroevolutionists, in this setting, should not be granted the privilege of presenting their view of this matter of origins as if it were thoroughly factual and as if its veracity were demonstrated beyond the shadow of a doubt. THIS POSITION IS SIMPLY NOT TRUE.

Wilbert H. Rusch, Sr.
2717 Cranbrook Rd.
Ann Arbor, MI 48104

XV. REFERENCES

[1]Giovanni Pettinato, "The Royal Archives of Tell Mardikh-Ebla," *Biblical Archeaologist*, XXXIX, 2 (May 1976), pp. 44-52.

[2]Clifford Wilson, *Rocks, Relics and Biblical Reliability*, Probe, Richardson, Texas, 1977, Chap. 13.

[3] William F. Albright, *Recent Discoveries in Bible Lands*, Funk & Wagnalls, NYC, 1955, pp. 70ff.

[4]Augustine, *City of God*, Modern Library, Random House, NYC, 1950, p. 406.

[5]Charles Darwin, *Origin of Species*, Everyman's Library, Dutton, NYC, 1971, p. 167.

[6]Wilbert Rusch, "Darwin's Last Hours," *Creation Research Society Quarterly*, XII, 2, pp. 99-102.

[7]David B. Kitts, "Paleontology and Evolutonary Theory," *Evolution*, XXVIII, September 1974, p. 467.

[8]Leonard Huxley, *Life and Letters of Thomas Henry Huxley*, Macmillan, NYC, Vol. 1, 1903, p. 180.

[9]Douglas Dewar and H. S. Shelton, *Is Evolution Proved?*, Hollis and Carter, London, 1947, p. 4.

[10]Aldous Huxley, "Ends and Means," quoted from R. F. R. Gardner, *Abortion, The Personal Dilemma*, Eerdmans, Grand Rapids, 1972, p. 57.

[11]*Encyclopedie Francaise*, Larousse, Paris, V, pp. 82-83.

[12]R.M. Cornelius, "Their Stage Drew All the World," *Tennessee Historical Quarterly*, XL, Summer 1981, pp. 1-15.

[13]*Evolution After Darwin*, Sol Tax, Ed., Un. of Chicago Press, Chicago, 1960, 3 vol.

[14]Norman Geisler, *The Creator in the Classroom*, Mott, Milford, Mich., 1982.

[15]Francis Bacon, *Advancement of Learning*, Colonial Press, NYC, 1900, p. 5.

[16]Alfred N. Whitehead, *Science and the Modern World*, Macmillan, NYC, 1926, p. 18.

[17]J. Robert Oppenheimer, "Science and Culture," *Encounter, Oct. 1946, p. 62.*

[18]Percy Raymond, *Bull. Geol. Soc. of America*, Vol. 46, p. 378.

[19]H. D. Pflug & H. Jaeschke-Boyer, "Combined Structural and Chemical Analysis of 3,800 Myr Old Microfossils," *Nature*, Vol. 280, pp. 483-6.

[20]Bartholomew Nagy, et al., "Amino Acids and the Hydrocarbons in 3,800 Myr. Old Rocks in the Isua Rocks, Southwestern Greenland,"

Nature, Vol. 289, 1/8 Jun. 1981, pp. 53-55.

[21] Wilbert H. Rusch, "Ontogeny Recapitulates Phylogeny,", *Creation Research Society Quarterly*, VI, 1, pp. 27ff.

[22] Theodosius Dobzhansky, Francisco Ayala, G. Ledyard Stebbins, and James W. Valentine, *Evolution*, Freeman, San Francisco, 1977, p. 8.

[23] Everett C. Olson and Jane Robinson, *Concepts of Evolution*, Merrill, Columbus, 1975, p. 10.

[24] Theodosius Dobzhansky, *Genetics and the Origin of Species*, 1941, 2nd ed., p. 7-8.

[25] Anthony Standen, *Science is a Sacred Cow*, Dutton, NYC, 1950 p. 101ff.

[26] G. H. Kerkut, *Implications of Evolution*, Pergamon, NYC, 1960, p. 157.

[27] Norman Macbeth, *Darwin Retried*. Gambit, Boston, 1971, pp. 4, 5.

[28] G. H. Kerkut, 1960, p. 157.

[29] Roger Lewin, "Evolutionary Theory Under Fire," *Science*, Vol. 210, Nov. 21, 1980, p. 883.

[30] U. S. Supreme Court Reports, 367 US 488, 6 L ed 2d 982, p. 987.

[31] Leonard Huxley, 1903, Vol. 1, p. 241.

[32] E. J. Corner, "Evolution," *Contemporary Botanical Thought*, McClead & Cobley, ed., Quadrangle, Chicago, 1961, p. 97.

[33] Marsh, Frank L., *Variation and Fixity in Nature*, Pacific Press, Mt. View, 1976, p. 36.

[34] Clark, Austin H., *The New Evolution*, Williams & Wilkins, Baltimore, 1930.

[35] Clark, Austin H., "Animal Evolution", *QUARTERLY REVIEW OF BIOLOGY*, December 1928.

[36] Willis, J. C., *The Course of Evolution*, Cambridge, London, 1940.

[37] Goldschmidt, Richard, *The Material Basis of Evolution*, Yale, New Haven, 1940.

[38] Nilsson, Heribert, *Synthetische Artbildung*, Gleerups, Lund, 1953, p. 1194-5.

[39] Ibid., p. 1186.

[40] Ibid., p. 1240-1.

[41] Ehrlich, Paul, and R. Holm, *The Process of Evolution* McGraw Hill, NYC, 1963, p. 66.

[42] Oakland Press, *Boston Baby Born with a Two Inch Tail*, Thursday, May 20, 1982.

[43] Clark, 1930, pp. 235-6.

[44] Lewin, 1980, pp. 883 ff.

[45] E. J. Corner, 1961, p. 97.
[46] Huxley, Thomas H., *Lay Sermons, Addresses, and Reviews*, Appleton, NYC, 1879, p. 225.
[47] Lewin, 1980, p. 883.
[48] Macbeth, 1971, pp. 124 ff.
[49] Himmelfarb, Gertrude, *Darwin and the Darwinian Revolution*, Chatto and Windus, London, 1959, pp. 310-311.
[50] *Life and Letters of Charles Darwin*, (3 Vol.), Murray, London, 1887, Vol. 2, p. 296.
[51] Darwin, 1971, p. 167.
[52] Stephen Jay Gould, *Ever Since Darwin*, Norton, NYC, 1977, p. 104.
[53] Wilbert H. Rusch, *Human Fossils*, Chap. 9, *Rock Strata and the Bible Record*, P. A. Zimmerman, Ed., Concordia, St. Louis, 1970, pp. 144-5.
[54] Rusch, *Ibid.*, pp. 156-159.
[55] Rusch, *Ibid.*, p. 154.
[56] McHenry, H. M.,"*A View of Hominid Lineage*," *Science*, Vol. 189 p.988.
[57] R. E. F. Leakey,"*Australopithecus, Homo erectus*, and the Single Species Hypothesis," *Nature*, Vol. 261, June 1976, pp. 572-5.
[58] Paul L. De Vore, Ed., *The Origin of Man*, Werner Gren Foundation, 1965, p. 11.
[59] Rusch, *Ibid*, pp. 171-2.
[60] Herbert Wray,"Lucy's Family Problems," *Science News*, Vol.124, July 2, 1983, pp. 8-11.
[61] Michael H. Day, "Lucy Jilted?", *Nature* Vol. 300, 16 December, 1982, p. 574.
[62] George Gaylord Simpson, "The Meaning of Taxonomic Statements," *Classification and Human Evolution*, Sherwood Washburn, Ed., Aldine, Chicago, 1963, p. 24.
[63] Dewar, Douglas, *The Transformist Illusion*, DeHoff, Murfreesboro, IN., 1957.
[64] Darwin, 1971, p. 442.
[65] Ibid., p. 294.
[66] Karl R. Popper, *Conjectures and Refutations*, Routledge & Kegan, London, 1963.
[67] John G. Funkhauser and John J. Naughton, "Radiogenic Helium and Argon in Ultramafic Inclusions from Hawaii," *Journal of Geophysical Research*, Vol. 73, 14, 1968 pp 4601-07.
[68] Dalrymple, G. A., and J. G. Moore, "Argon 40: Excess in Submarine Pillow Basalts from Kilauea Volcano, Hawaii," *Science*, Vol. 161, pp. 1132-1135.

[69]Barnes, Thomas G., *Origin and Destiny of the Earth's Magnetic Field,* Tech. Mon. #4, ICR, San Diego, 1973.

[70]Gentry, Robert V., et al, "Radiohalos in a Radio-Chronological and Cosmological Perspective," *Science,* 5 April, 1974, Vol. 184, pp. 62-66.

[71]Robert Gentry, "Radiohalos in Coalified Wood: New Evidence Relating to the Time of Uranium Introduction and Coalification", Science," Vol. 194, 15 Oct. 1976, pp. 315-8.

[72]Robert Gentry, "Radioactive Halos," *Annual Review of Nuclear Science,* XXIII, 1973.

[73]Harold F. Blum, *Time's Arrow and Evolution,* Princeton Un. Press, Princeton, 1951, p. 10.

[74]James C. Hefley, *Are Textbooks Harming Your Children?,* Mott Media, Milford, 1979, pp. 43-51.

[75]Macbeth, 1971, p. 124 ff.

[76]W. R. Thompson, "Introduction," *Origin of Species,* Dutton, NYC, 1956, p. XXIII.

APPENDIX A.
BIBLIOGRAPHY

A word with regard to the general bibliography in this appendix. This list of books deals with some aspect or other of the argument on origins. Most expand the arguments pro and con that appear in the main work. Examples would be #7, 12, 14, 15, 18, 20, 21, and 22. Others deal with education and school questions, e.g. #8 and 9. Still others deal with related archeological concerns, e.g. #19. An interesting fact is that a number of the books listed here are at times found in the bibliographies of those secular texts that at least mention the subject of creation. Examples would be *The Science Of Evolution,* by Stanfield and *The World Of Life,* by Salomon and Davis.

The works listed in this bibliography can be classed as scholarly i.e. they are fully and accurately footnoted. The references found in them and which are quoted, are taken from reputable works, written by individuals knowledgeable in their respective fields. Therefore the writer has no hesitancy in recommending these books to the reader who may wish to delve further into this whole area for additional information. However, this is not to imply that these are the only works on the subject of origins that are worthwhile and reputable.

1. Anderson, J. Kerby & Harold G. Coffin, *Fossils In Focus,* Probe Ministries, Dallas, 1977.

2. Aw, S. E., *Chemical Evolution,* Mott Media, Milford, MI, 1982.

3. Coppedge, James E., *Evolution: Possible or Impossible,* Zondervan, Grand Rapids, 1973.

4. Conn, Harry, *Four Trojan Horses,* Mott Media, Milford, MI, 1982.

5a. Darwin, Charles, *Origin of Species,* E. P. Dutton and Co., NYC, 1956. N. B. This edition is the one that contains Dr. W. R. Thompson's introduction.

5b. Darwin, Charles, *Origin of Species,* E. P. Dutton and Co., NYC, 1974. N. B. This edition has the foreword by L. Harrison Matthews.

6. Davis, P. William and Eldra Solomon, *The World of Biology,* McGraw-Hill, NYC, 1974. N. B. This is a college biology text that at least recognized the existence of creation as an alternate concept.

7. Frair, Wayne and P. W. Davis, *A Case For Creation,* 3rd edition, Moody, Chicago, 1983.

8. Geisler, Norman, *The Creator in the Courtroom,* Mott Media, Milford, MI, 1982.

9. Hefly, James C., *Are Textbooks Harming Your Children?*, Mott Media, Milford, Mich., 1979.

10. Hoyle, Frederick, and N. C. Wickramasinghe, *Evolution From Space*, Dent, London, 1981.

11. Kerkut, G. A., *Implications of Evolution*, Pergamon, NYC, 1960. N. B. The average reader can follow chapters 1, 2 and 10.

12. Klotz, John, *Genes, Genesis and Evolution*, 2nd edition, Concordia, St. Louis, 1970.

13. Lammerts, Walter, editor, *Why Not Creation*, Creation Research Society Books, Norcross, Ga., 1970.

14. Macbeth, Norman, *Darwin Retried*, Gambit, Boston, 1971.

15. Marsh, Frank, *Life, Man, and Time*, Outdoor Pictures, Anacortes, 1967.

16. Marsh, Frank, *Variation and Fixity in Nature*, Creation Research Society Books, Norcross, 1976.

17. Morris, Henry and Gary Parker, *What Is Creation Science?*, Creation Life, San Diego, 1982.

18. Whitcomb, John and Henry Morris, *The Genesis Flood*, Presbyterian, Nutley, 1961.

19. Wilson, Clifford, *Rocks, Relics, and Biblical Reliability*, Probe Ministries, Dallas, 1977.

20. Wysong, R. L. *The Creation-Evolution Controversy*, Inquiry Press, Midland, 1976.

21. Zimmerman, Paul A., editor, *Creation, Evolution, and God's Word*, Concordia, St. Louis, 1972.

22. Zimmerman, Paul A., editor, *Darwin, Evolution, and Creation*, Concordia, St. Louis, 1959.

23. Zimmerman, Paul A., editor, *Rock Strata and the Bible Record*, Concordia, St. Louis, 1970.

APPENDIX B
QUOTES

A further word about the quotes included in this appendix; you will note that they are all by evolutionists. This may be somewhat surprising, considering what they are saying, but the authors are candid men and tell it like it is. The writer quotes them because they know what they are talking about, but also because they certainly could be depended on to minimize any adverse evidence. All of the quotes are given in their entireties, not out of context, not segmented.

"Great is the power of steady misinterpretation".
Charles Darwin, *The Origin of Species,* 6th ed. 1872, Burt, p. 395.

"Much evidence can be in favor of the theory of evolution—from biology, bio-geography and paleontology, but I still think that to the unprejudiced, the fossil record of the plants is in favor of special creation. If, however, another explanation could be found for this hierarchy of classification, it would be the knell of the theory of evolution. Can you imagine how an orchid, a duckweed, and a palm have come from the same ancestry, and have we any evidence for this assumption? The evolutionist must be prepared with an answer, but I think that most would break down before an inquisition."
E. J. Corner, Evolution, from *Contemporary Botanical Thought,* A. M. MacLeod and L. S. Cobley, editors, Quadrangle, Chicago, 1961, p. 97.

"There is a theory which states that many living animals can be observed over the course of time to undergo changes so that new species are formed. This can be called the 'Special Theory of Evolution' and can be demonstrated in certain cases by experiments. On the other hand there is the theory that all living forms in the world have arisen from a single source which itself came from an inorganic form. This theory can be called the 'General Theory of Evolution' and the evidence that supports it is not sufficiently strong to allow us to consider it as anything more than a working hypothesis. It is not clear whether the changes that bring about speciation are of the same nature as those that brought about the development of new phyla. The answer will be found by future experimental work and not by dogmatic assertions that the General Theory of Evolution must be correct because there is nothing else that will satisfactorily take its place."

G. A. Kerkut, *Implications of Evolution,* Pergamon, NYC, 1960, p. 157.

"Much of Professor Thompson's criticism of Darwin's text is unanswerable. In accepting evolution as a fact, how many biologists pause to reflect that science is built upon theories that have been proved by experiment to be correct, or remember that the theory of animal evolution has never been thus proved? Even 'Darwin's Bulldog' as Thomas Huxley once called himself, wrote in 1863; 'I adopt Mr. Darwin's hypothesis, therefore, subject to the production of proof that physiological species may be produced by selective breeding'—meaning species that are fertile if crossed. That proof has never been produced, though a few not-entirely-convincing examples are claimed to have been found. The fact of evolution is the backbone of biology, and biology is thus in the peculiar position of being a science founded on an unproved theory—is it then a science or a faith? Belief in the theory of evolution is thus exactly parallel to belief in special creation—both are concepts which believers know to be true but neither, up to the present, has been capable of proof."

L. Harrison Matthews, Introduction to *Origin of Species,* Dutton, NYC, 1974, pp. x-xi.

"It is very depressing to find that many subjects are becoming encased in scientific dogmatism. The basic information is frequently overlooked or ignored and opinions become repeated so often and so loudly that they take on the tone of Laws."

G. A. Kerkut, *Implications of Evolution,* p. viii (1960).

"The demonstration that the idea of special creation is scientifically meaningless does not however, 'prove' that the theory of evolution is correct. Current faith in the theory is reminiscent of many other ideas which at one time were thought to be self-evidently true and supported by all available data . . . . Perpetuation of today's theory as dogma will not encourage progress toward more satisfactory explanations of observed phenomena."

Ehrlich and Holm, *The Process of Evolution,* p. 310 (1963).

"But a much deeper and more penetrating analysis of the problem as put together by Professor Ronald H. Brady of Ramapo College in the quarterly called *Systematic Zoology* for December 1979. He has about a 21-page article on natural selection and criteria for judging it. He seemed to me to utterly destroy the entire idea of natural selection as presently conceived. He does it in a philosophical and scientific way, to which there has been as yet, as far as I'm aware, no serious reply. I think it destroys the idea of natural selection, and this is certainly the opinion of many people at the American Museum of Natural History. The whole basis for the Synthetic Theory is shot to pieces right there in his article."

Norman Macbeth in an interview reported in the Spring 1982 issue of *Towards*.

APPENDIX C
CREATIONIST ORGANIZATIONS

United States

1. Baltimore Creation Fellowship
   P.O. Box 356
   Perry Hall, MD 21236
   (c, e)

2. Bible-Science Association
   2911 E. 42nd St.
   Minneapolis, MN 55406
   (b, c, d, e)

3. Center for Scientific Creation
   1319 Brush Hill Circle
   Naperville, IL 60540

4. Citizens for Fairness in Education
   953 Longhorn Drive
   Plano, TX 75023
   (f)

5. Citizens for Scientific Creation
   P.O.Box 164
   Saratoga, CA 85070

6. Creation Health Foundation
   19 Gallery Centre
   Taylors, SC 29687
   (d)

7. Creation Research Society
   P.O. Box 14016
   Terre Haute, IN 47803
   (a)

8. Creation Science Association
   2825 Riva Ridge
   Cottage Grove, WI 53527
   (a, c, d)

9. Creation Science Association
   18346 Beverly Road
   Birmingham, MI 48009
   (d)

10. Creation Science Association of Orange County
    P.O. Box 4325
    Irvine, CA 92716-4325
    (b)

11. Creation Science Association of Michigan
    26938 Northmore
    Dearborn, MI 48127
    (d)

12. Creation Science Committee
    P.O. Box 2282
    Fairbanks, AK 99707

13. Creation-Science Society of Milwaukee
    5334 N. 66th St.
    Milwaukee, WI 53218

14. Creation Science Society
    201 S. Brent St.
    Ventura, CA 93003

15. Creation Science Research Center
    P.O. Box 23195
    San Diego, CA 92123
    (b, c, d, e, f)

16. Creation Social Science & Humanities Society
    1429 Holyoke
    Wichita, KS 67208
    (a)

17. Fair Education Foundation
    Rte. #2, Box 415
    Murphy, NC 28906
    (b, e)

18. Geoscience Research Institute
    Loma Linda Univ.
    Loma Linda, CA 92350
    (a)

19. Institute for Creation Research
    2100 Greenfield Dr.
    El Cajon, CA 92021
    (b, c, e)

20. Lutheran Research Society
    2222 B. Street
    Forest Grove, OR 97116

21. Lutheran Science Institute
    8830 W. Bluemound Rd.
    Milwaukee, WI 53226
    (d)

22. Missouri Association for Creation, Inc.
    P.O. Box 23984
    St. Louis, MO 63119
    (b, d, e)

23. National Foundation for Fairness in Education
    P.O. Box 1
    Washington, DC 20044
    (b, e)

24. Students for Origins Research
    P.O. Box 203
    Goleta, CA 93116
    (d, f)

25. Twin Cities Creation Science Association
    2945 Van Nest Ave. S
    Minneapolis, MN 55409
    (f)

CANADA

1. Creation Science Association of Ontario
   P.O. Box 821, Stat. A,
   Scarborough, Ontario
   CANADA, M1K 5C8
   (d)

2. Creation Science Association of Canada
   P.O. Box 34006
   Vancouver, B.C.,
   CANADA V6J 4M1
   (b, c, e, f)

3. Creation Science Association of Manitoba
   P.O. Box 68
   Rosenort, Manitoba
   CANADA

4. Creation Science Association of Alberta
   P.O. Box 9075 Stat. E.
   Edmonton, Alberta
   CANADA T5P 4K1

5. Creation Science of Saskatchewan
   P.O. Box 1821
   Prince Albert, Sask.
   CANADA S6V 6J9

6. International Christian Crusade
   205 Yonge St. Room 31
   Toronto,
   CANADA V8N 1C5
   (b, d)

7. North America Creation Movement
   P.O. Box 5083, Stat. "B"
   Victoria, B.C.
   CANADA V8R 6N3
   (b, d, e)

AUSTRALIA

1. Creation-Science Foundation, Limited
   P.O. Box 302
   Sunnybank, Queensland
   AUSTRALIA 4109 (a, b)

2. Creation-Science Foundation
   Cnr. Bradman & Bellrick
   Acacia Ridge, Queensland
   AUSTRALIA 4110 (a, e)

3. Creation-Science Movement
   "Bethuel" 13 Beddows St.
   Birwood, Victoria
   AUSTRALIA 3125

UNITED KINGDOM

1. Biblical Creation Society
   51 Cloan Crescent
   Bishopbriggs, Glasgow,
   SCOTLAND G64 2 Hn
      (a, b)

2. Creation News
   3 Church Terrace
   Cardiff,
   WALES CF2 5 AW
      (b, d)

3. Creation-Science Movement
   Rivendell 20 Foxley Lane
   High Salvington, Worthing
   ENGLAND BN13 3Ab
      (a, b)

4. Creation-Science Movement
   13 Argyle Ave.
   Hounslow, Middlesex
   ENGLAND TW3 2LE

5. Research Scientists Christian Fellowship
   39 Bedford Square
   London, ENGLAND
   WC1 B 3EY

6. Research Scientists Christian Fellowship
   38 De Montfort St.
   Leicester,
   ENGLAND LEI 7GP

7 Somerset Creation Group
   Mead Farm, Downhead,
   West Camel
   Yeovil, Somerset
   ENGLAND BA22 7RQ
   (d)

INDIA

1. Creation Research and Communication Project
   Prathana Bhaven, 14 Mahavir Colony,
   Naya Bazar, Gwalior, M.P. India 474009 (a, b)

KOREA

1. Korea Association of Creation Research
   15-5 Jung-Dong, Jung-Ku., Seoul 100, KOREA (d, e)

NETHERLANDS

1. Stitchting tot Bevordering van Bybelgetrouwe
   Wetenschap (Foundation for the Advancement of Studies
   Faithful to the Bible) Secretary: Posbus G-57
   3800 AZ Amersfoort, NETHERLANDS (a, b, e)

NEW ZEALAND

1. Creation Literature Society, Inc
   48 Craig Road, Maraetai Beach, Auckland
   NEW ZEALAND

SWEDEN

1. Association for Christian Belief
   Okome, Prastgard P1, 4703
   31060 Ullard SWEDEN

2. Forening For Biblisk Skapelesetro
   Box 3170, 400 10 Goteborg
   SWEDEN (b, e, a)

SPAIN

1. Coordinadora Creacionista
   Apartado 20022
   Sabadell, (Barcelona) SPAIN

PUBLISHERS

1. Beta Books
   P.O. Box 23605
   San Diego, CA 92123

2. Bob Jones University Press
   Greenville, SC 29614

3. Concordia Publishing House
   3558 S. Jefferson
   St. Louis, MO 63118

4. Creation-Life Publishers
   P.O. Box 15666
   San Diego, CA 92115

5. Creation Research Books
   5093 Williamsport Dr.
   Norcross, GA 30092

6. Mott Media
   Milford, MI 48042

7. Probe Ministries
   12011 Colt Rd. Ste. 107
   Dallas, TX 75251

AV SUPPLIERS

1. Creation Filmstrip Center
   Route #1
   Haviland, KS 67059

2. Creation Science Educational Media
   P.O. Box 302
   Sunnybank, Queensland
   AUSTRALIA 4109

3. Films for Christ Association
   5310 N. Eden Road
   Elmwood, IL 61529

4. CLP Video
   1250 Fayette St.
   El Cajon, CA 92020

5. Crossroads Christian Communications
   100 Huntley St.
   Toronto, Ontario,
   CANADA M4Y 2L1

Code: (a) journal
      (b) pamphlets, tracts
      (c) filmstrips, cassettes
      (d) newsletter
      (e) books
      (f) other selected

APPENDIX D
THE HUMANIST MANIFESTO II

The next century can be and should be the humanist century. Dramatic scientific, technological, and ever-accelerating social and political changes crowd our awareness. We have virtually conquered the planet, explored the moon, overcome the natural limits of travel and communication; we stand at the dawn of a new age, ready to move farther into space and perhaps inhabit other planets. Using technology wisely, we can control our environment, conquer poverty, markedly reduce disease, extend our lifespan, significantly modify our behavior, alter the course of human evolution and cultural development, unlock vast new powers, and provide humankind with unparalleled opportunity for achieving an abundant and meaningful life.

The future is, however, filled with dangers. In learning to apply the scientific method to nature and human life, we have opened the door to ecological damage, overpopulation, dehumanizing institutions, totalitarian repression, and nuclear and biochemical disaster. Faced with apocalyptic prophesies and doomsday scenarios, many flee in despair from reason and embrace irrational cults and theologies of withdrawal and retreat.

Traditional moral codes and new irrational cults both fail to meet the pressing needs of today and tomorrow. False "theologies of hope" and messianic ideologies, substituting new dogmas for old, cannot cope with existing world realities. They separate rather than unite peoples.

Humanity, to survive, requires bold and daring measures. We need to extend the uses of scientific method, not renounce them, to fuse reason with compassion in order to build constructive social and moral values. Confronted by many possible futures, we must decide which to pursue. The ultimate goal should be the fulfillment of the potential for growth in each human personality not for the favored few, but for all of human kind. Only a shared world and global measures will suffice.

A humanist outlook will tap the creativity of each human being and provide the vision and courage for us to work together. This outlook emphasizes the role human beings can play in their own spheres of action. The decades ahead call for dedicated, clear-minded men and women able to marshall the will, intelligence, and cooperative skills for shaping a desirable future. Humanism can provide the purpose and inspiration that so many seek; it can give personal meaning and significance to human life.

Many kinds of humanism exist in the contemporary world. The varieties and emphases of naturalistic humanism include "scientific," "ethical," "democratic," "religious," and "Marxist" humanism. Free thought, atheism, agnosticism, skepticism, deism, rationalism, ethical culture, and liberal religion all claim to be heir to the humanism tradition. Humanism traces its roots from ancient China, classical Greece and Rome, through the Renaissance and the Enlightenment, to the scientific revolution of the modern world. But views that merely reject theism are not equivalent to humanism. They lack commitment to the positive belief in the possibilities of human progress and to the values central to it. Many within religious groups, believing in the future of humanism, now claim humanistic credentials. Humanism is an ethical process through which we all can move, above and beyond the divisive particulars, heroic personalities, dogmatic creeds, and ritual customs of past religions or their mere negation.

We affirm a set of common principles that can serve as a basis for united action-positive principles relevant to the present human condition. They are a design for a secular society on a planetary scale.

For these reasons, we submit this new *Humanist Manifesto* for the future of humankind; for us, it is a vision of hope, a direction for satisfying survival.

RELIGION

First: In the best sense, religion may inspire dedication to the highest ethical ideals. The cultivation of moral devotion and creative imagination is an expression of genuine "spiritual" experience and aspiration.

We believe, however, that traditional dogmatic or authoritarian religions that place revelation, God, ritual, or creed above human needs and experience do a disservice to the human species. Any account of nature should pass the tests of scientific evidence; in our judgment, the dogmas and myths of traditional religions do not do so. Even at this late date in history, certain elementary facts based upon the critical use of scientific reason have to be restated. We find insufficient evidence for belief in the existence of a supernatural; it is either meaningless or irrelevant to the question of the survival and fulfillment of the human race. As nontheists, we begin with humans not God, nature not deity. Nature may indeed be broader and deeper than we now know; any new discoveries, however, will but enlarge our knowledge of the natural.

Some humanists believe we should reinterpret traditional religions and reinvest them with meanings appropriate to the current situation. Such redefinitions, however, often perpetuate old dependencies and escapisms; they easily become obscurantist, impeding the free use of the intellect. We need, instead, radically new human purposes and goals.

We appreciate the need to preserve the best ethical teachings in the religious traditions of humankind, many of which we share in common.

But we reject those features of traditional religious morality that deny humans a full appreciation of their own potentialities and responsibilities. Traditional religions often offer solace to humans, but, as often, they inhibit humans from helping themselves, or experiencing their full potentialities. Such institutions, creeds, and rituals often impede the will to serve others. Too often traditional faiths encourage dependence rather than independence, obedience rather than affirmation, fear rather than courage. More recently they have generated concerned social action, with many signs of relevance appearing in the wake of the "God is Dead" theologies. But we can discover no divine purpose or providence for the human species. While there is much that we do not know, humans are responsible for what we are or will become. No deity will save us; we must save ourselves.

Second: Promises of immortal salvation or fear of eternal damnation are both illusory and harmful. They distract humans from present concerns, from self-actualization, and from rectifying social injustices. Modern science discredits such historic concepts as the "ghost in the machine" and the "separable soul". Rather, science affirms that the human species is an emergence from natural evolutionary forces. As far as we know, the total personality is a function of the biological organism transacting in a social and cultural context. There is no creditable evidence that life survives the death of the body. We continue to exist in our progeny and in the way that our lives have influenced others in our culture.

Traditional religions are surely not the only obstacles to human progress. Other ideologies also impede human advance. Some forms of political doctrine, for instance, function religiously, reflecting the worst features of orthodoxy and authoritarianism, especially when they sacrifice individuals on the altar of utopian promises. Purely economic and political viewpoints, whether capitalistic or communist, often function as religious and ideological dogma. Although humans undoubtedly need economic and political goals, they also need creative values by which to live.

ETHICS

Third: We affirm that moral values derive their source from human experience. Ethics are *autonomous* and *situational,* needing no theological or ideological sanction. Ethics stem from human need and interest. To deny this distorts the whole basis of life. Human life has meaning because we create and develop our futures. Happiness and the creative realization of human needs and desires, individually and in shared enjoyment, are continuous themes of humanism. We strive for the good life, here and now. The goal is to pursue life's enrichment, despite debasing forces of vulgarization, commercialization, bureaucratization, and dehumanization.

Fourth: Reason *and intelligence* are the most effective instruments that humankind possesses. There is no substitute: neither faith nor passion suffices in itself. The controlled use of scentific methods, which have transformed the natural and social sciences since the Renaissance, must be extended further in the solution of human problems. But reason must be tempered by humility, since no group has a monopoly of wisdom or virtue. Nor is there any guarantee that all problems can be solved or all questions answered. Yet critical intelligence, infused by a sense of human caring, is the best method that humanity has for resolving problems. Reason should be balanced with compassion and empathy and the whole person fulfilled. Thus, we are not advocating the use of scientific intelligence independent of or in opposition to emotion, for we believe in the cultivation of feeling and love. As science pushes back the boundary of the known, man's sense of wonder is continually renewed, and art, poetry, and music find their places, along with religion and ethics.

THE INDIVIDUAL

Fifth: The preciousness and dignity of the individual person is a central humanist value. Individuals should be encouraged to realize their own creative talents and desires. We reject all religious, ideological, or moral codes that denigrate the individual, suppress freedom, dull intellect, dehumanize personality. We believe in maximum individual autonomy consonant with social responsibility. Although science can account for the causes of behavior, the possibilities of individual *freedom of choice* exist in human life and should be increased.

Sixth: In the area of sexuality, we believe that intolerant attitudes, often cultivated by orthodox religions and puritanical cultures, unduly repress sexual conduct. The right to birth control, abortion, and divorce should be recognized. While we do not approve of exploitive, denigrating forms of sexual expression, neither do we wish to prohibit, by law or social sanction, sexual behavior between consenting adults. The many varieties of sexual exploration should not in themselves be considered "evil". Without countenancing mindless permissiveness or unbridled promiscuity, a civilized society should be a tolerant one.

DEMOCRATIC SOCIETY

Seventh: To enhance personal freedom and dignity the individual must experience a full range of *civil liberties* in all societies. This includes freedom of speech and the press, political democracy, the legal right of opposition to governmental policies, fair judicial process, religious liberty, freedom of association, and artistic, scientific, and cultural freedom. It also includes a recognition of an individual's right to die with dignity, euthanasia, and the right to suicide. We oppose the increasing invasion of privacy, by whatever means, in both totalitarian and democratic societies. We would safeguard, extend and implement the principles of human freedom evolved from the *Magna Carta* to the *Rights of Man,* and the *Universal Declaration of Human Rights.*

Eighth: We are committed to an open and democratic society. We must extend *participation democracy* in its true sense to the economy, the school, the family, the workplace, and voluntary associations. Decision-making must be decentralized to include widespread involvement of people at all levels — social, political, and economic. All persons should have a voice in developing the values and goals that determine their lives. Institutions should be responsive to expressed desires and needs. The conditions of work, education, devotion, and play should be humanized. Alienating forces should be modified or eradicated and bureaucratic structures should be held to a minimum. People are more important than decalogues, rules, proscriptions, or regulations.

Ninth: The separation of church and state and the separation of ideology and state are imperatives. The state should encourage maximum freedom for different moral, political, religious, and social values in society. It should not favor any particular religious bodies through the use of public monies, nor espouse a single ideology and function thereby as an instrument of propaganda or oppression, particularly against dissenters.

Tenth: Human societies should evaluate economic systems not by rhetoric or ideology, but by whether or not they *increase economic well being* for all individuals and groups, minimize poverty and hardship, increase the sum of human satisfaction, and enhance the quality of life. Hence the door is open to alternative economic systems. We need to democratize the economy and judge it by its responsiveness to human needs, testing results in terms of the common good.

Eleventh: The principle of moral equality must be furthered through the elimination of all discrimination based upon race, religion, sex, age, or national origin. This means equality of opportunity and recognition of talent and merit. Individuals should be encouraged to contribute to their own betterment. If unable, then society should provide the means to satisfy their basic economic, health and cultural needs, including wherever resources make possible, a minimum guaranteed annual income. We are concerned for the welfare of the aged, the infirm, the disadvantaged, and also for the outcasts — the mentally retarded, abandoned or abused children, the handicapped, prisoners, addicts —

for *all* who are neglected or ignored by society. Practicing humanists should make it their vocation to humanize personal relations.

We believe in the *right to universal education.* Everyone has a right to the cultural opportunity to fulfill his or her unique capacities and talents. The schools should foster satisfying and productive living. They should be open at all levels to any and all; the achievement of excellence should be encouraged. Innovative and experimental forms of education are to be welcomed. The energy and idealism of the young deserve to be appreciated and channeled to constructive purposes.

We deplore racial, religious, ethnic, or class antagonisms. Although we believe in cultural diversity and encourage racial and ethnic pride, we reject separations which promote alienation and set people and groups against each other; we envision an *integrated* community where people have a maximum opportunity for free and voluntary association.

We are *critical of sexism or sexual chauvinism* — male or female. We believe in equal rights for both men and women to fulfill their unique careers and potentialities as they see fit, free of invidious discrimination.

WORLD COMMUNITY

Twelfth: We deplore the division of humankind on nationalistic grounds. We have reached a turning point in human history where the best option is to *transcend the limits of national sovereignty* and to move toward the building of a world community in which all sectors of the human family can participate. Thus we look to the development of a system of world law and a world order based upon transnational federal government. This would appreciate cultural pluralism and diversity. It would not exclude pride in national origins and accomplishments nor the handling of regional problems on a regional basis. Human progress, however, can no longer be achieved by focusing on one section of the world, Western or Eastern, developed or underdeveloped. For the first time in human history, no part of humankind can be isolated from any other. Each person's future is in some way linked to all. We thus reaffirm a commitment to the building of a world community, at the same time recognizing that this commits us to some hard choices.

Thirteenth: This world community must *renounce the resort to violence and force* as a method of solving international disputes. We believe in the peaceful adjudication of differences by international courts and by the development of the arts of negotiation and compromise. War is obsolete. So is the use of nuclear, biological, and chemical weapons. It is a planetary imperative to reduce the level of military expenditures and turn these savings to peaceful and people-orientated uses.

Fourteenth: The world community must engage in *cooperative planning* concerning the use of rapidly depleting resources. The planet Earth must be considered a single *ecosystem*. Ecological damage, resource depletion, and excessive population growth must be checked by international concord. The cultivation and conservation of nature is a moral value; we should perceive ourselves as integral to the sources of our being in nature. We must free our world from needless pollution and waste, responsibly guarding and creating wealth, both natural and human. Exploitation of natural resources, uncurbed by social conscience, must end.

Fifteenth: The problems of *economic growth and development* can no longer be resolved by one nation alone; they are worldwide in scope. It is the moral obligation of the developed nations to provide — through an international authority that safeguards human rights — massive technical, agricultural, medical, and economic assistance, including birth-control techniques, to the developing portions of the globe. World poverty must cease. Hence, extreme disproportions in wealth, income, and economic growth should be reduced on a worldwide basis.

Sixteenth: Technology is a vital key to human progress and development. We deplore any necromantic efforts to condemn indiscriminately all technology and science or to counsel retreat from its further extension and use for the good of humankind. We would resist any moves to censor basic scientific research on moral, political, or social grounds. Technology must, however, be carefully judged by the consequences of its use; harmful and destructive changes should be avoided. We are particularly disturbed when technology and bureaucracy control, manipulate, or modify human beings without their consent. Technological feasibility does not imply social or cultural desirability.

Seventeenth: We must expand communication and transportation across frontiers. Travel restriction must cease. The world must be open to diverse political, ideological, and moral viewpoints and evolve a worldwide system of television and radio for information and education. We thus call for full international cooperation in culture, science, the arts, and technology *across ideological borders.* We must learn to live openly together or we shall perish together.

HUMANITY AS A WHOLE

In closing: The world cannot wait for a reconciliation of competing political or economic systems to solve its problems. These are the times for men and women of good will to further the building of a peaceful and prosperous world. We urge that parochial loyalties and inflexible moral and religious ideologies be transcended. We urge recognition of the common humanity of all people. We further urge the use of reason and compassion to produce the kind of world we want — a world in which peace, prosperity, freedom, and happiness are widely shared. Let us not abandon that vision in despair or cowardice. We are responsible for what we are or will be. Let us work together for a humane world by means commensurate with humane ends. Destructive ideological differences among communism, capitalism, socialism, conservatism, liberalism, and radicalism should be overcome. Let us call for an end to terror and hatred. We will survive and prosper only in a world of shared humane values. We can initiate new directions for humankind: ancient rivalries can be superseded by broad-based cooperative efforts. The commitment to tolerance, understanding, and peaceful negotiation does not necessitate acquiescence to the status quo nor the damming up of dynamic and revolutionary forces. The true revolution is occurring and can continue in countless non-violent adjustments. But this entails the willingness to step forward onto new and expanding plateaus. At the present juncture of history, commitment to all humankind is the highest commitment of which we are capable; it transcends the narrow allegiances of church, state, party, class, or race in moving toward a wider vision of human potentiality. What more daring a goal for humankind than for each person to become, in ideal as well as in practice, a citizen of a world community. It is a classical vision; we can now give it new vitality. Humanism thus interpreted is a moral force that has time on its side. We believe that humankind has the potential intelligence, good will, and cooperative skill to implement this commitment in the decades ahead.

We, the undersigned, while not necesarily endorsing every detail of the above, pledge our general support to *Humanist Manifesto II* for the future of mankind. These affirmations are not a final credo or dogma but an expression of a living and growing faith. We invite others in all lands to join us in further developing and working for these goals.

This article first appeared in *The Humanist* issue of Sept./Oct. 1973, Volume XXXIII No. 5, and is reprinted by permission.

APPENDIX E

A Statement Affirming Evolution as a Principle of Science

For many years it has been well established scientifically that all known forms of life, including human beings, have developed by a lengthy process of evolution. It is also verifiable today that very primitive forms of life, ancestral to all living forms, came into being thousands of millions of years ago. They constitute the trunk of a "tree of life" that, in growing, branched more and more; that is, some of the later descendents of these earliest living things, in growing more complex, became ever more diverse and increasingly different from one another. Humans and the other highly organized types of today constitute the present twig-end of that tree. The human twig and that of the apes sprang from the same apelike progenitor branch.

Scientists consider that none of their principles, no matter how seemingly firmly established — and no ordinary "facts" of direct observation either — are absolute certainties. Some possibility of human error, even if very slight, always exists. Scientists welcome the challenge of further testing of any view whatever. They use such terms as *firmly established* only for conclusions founded on rigorous evidence that have continued to withstand searching criticism.

The principle of biological evolution, as just stated, meets these criteria exceptionally well. It rests upon a multitude of discoveries of very different kinds that concur and complement one another. It is therefore accepted into humanity's general body of knowledge by scientists and other reasonable persons who have familiarized themselves with the evidence.

In recent years, the evidence for the principle of evolution has continued to accumulate. This has resulted in a firm understanding of biological evolution, including the further confirmation of the principle of natural selection and adaptation that Darwin and Wallace over a century ago, showed to be an essential part of the process of biological evolution.

There are no alternative theories to the principle of evolution, with its "tree of life" pattern, that any competent biologist of today takes seriously. Moreover, the principle is so important for an understanding of the world we live in and of ourselves that the public in general, including students taking biology in school, should be made aware of it, and of the fact that it is firmly established in the view of the modern scientific community.

Creationism is not scientific, it is a purely religious view, held by some religious sects and persons and strongly opposed by other religious sects and persons. Evolution is the only presently known strictly scientific and nonreligious explanation for the existence and diversity of living organisms. It is therefore the only view that should be expounded in public-school courses on science, which are distinct from those on religion.

We, the undersigned, call upon all local school boards, manufacturers of textbooks and teaching materials, elementary and secondary teachers of biological science, concerned citizens, and educational agencies to do the following:

—Resist and oppose measures currently before several state legislatures that would require creationist views of origins be given equal treatment and emphasis in public-school biology classes and text materials,

—Reject the concept currently being put forth by certain religious and creationist pressure groups that alleges that evolution is itself a tenet of a religion of "secular humanism", and as such is unsuitable for inclusion in the public school science curriculum,

—Give vigorous support and aid to those classroom teachers who present the subject matter of evolution fairly and who often encounter community opposition.

This article first appeared in *The Humanist* issue of Jan./Feb. 1977, and is reprinted by permission.

INDEX

Australopithecus 46, 47-9
age of earth 50 ff.
Aquinas, Thomas 2
Augustine 2
Bacon, Francis 15-7
baramin 31
BSCS biology texts 12-4
Cambrian period 26
Clark, Austin A. 32-3
continuum of life 26, 32
creation-science 15
creation, beginning of 1
creation, definition 26
creation, religion and 27
creationist scientists 16
Cuvier, Georges 4
Darwin Centennial 12
Darwin, Charles 6, 36, 40
Darwinism, definition of 6
Davies, L. Merson 50
dating, radiogenic
  anomalies 52
  assumptions 51
  extrapolation 53
  pillow lavas 52
Dewar, Douglas 50
embryology 37
evolution and *Augustine* 2
evolution, beginning of 1
evolution, definition of 23
evolution-science 15
geographic distribution 6, 36
Goldschmidt, Richard B. 33
Gould, Stephen 41
Haeckel, Ernst 37
hopeful monster 34
human characteristics 43
human variation 42
humanity, criteria for 45
Huxley, Alduous 10
Huxley, Thomas H. 9, 39
hypothesis 18
Jenkin, Fleeming 33
Kerkut, G. A. 24

kind 3
Kitts, David 7
Lamarck, Jean de 3ff.
LeMoine, Paul 11
Linnaeus, 3, 30
Leakey, Richard 47-8
'Lucy' 48
macroevolution 32, 35, 37
magnetic field 53
Marx, Karl 11
microevolution 35
monophyletic 33
Muller, H. J. 12
Nilsson, Heribert 34
Paley, William 5, 40
parents, dilemma of 56
paleontology 39
Piltdown Hoax 40
pleochroic halos 54
pre-Cambrian fossils 17ff
phylogenies 34
polyphyletic 33
reconstructions, human 45
science, definition of 17 ff
science and origins 21
scientific inquiry 19
science, speculations in 19ff.
Scopes, John T. 11
Shaw, George Bernard 9
Sivaithecus 45
Skhul 45
Simpson, George Gaylord 48
species, definition of 30
species, fixity of 36
symbiosis 15
time 51
transitional forms 39
uniformitarianism 22
uniformity 22
vestigial organs 37-8
variation 30
Whitehead, Alfred North 17
Willis, J. C. 33